Detailed Critical Review of:

The Hunting Apes

Meat Eating and the Origins of Human Behavior,

by Professor Craig Stanford, U South. Calif.

and,

Catching Fire

AF824763

How Cooking Made us Human

by Professor Richard Wrangham, Harvard U

First e-editions 2010, this 1st Edition 2011 now with responses from right of reply offers to Prof. Stanford and Prof. Wrangham as at July 2011.

David J Vance, Multifactor Health and Education Initiative:

BSc (med physiol, biochem, chem) (Hons: nutrition), James Cook U
BA (Psychology), James Cook University
Master of Medical Science (Epidemiology), University of Queensland
Master of Public Health (Nutrition), University of Queensland
Master of Medical Statistics, University of Newcastle,
PGrad Diploma of Public Health & Tropical Medicine, James Cook U
Grad Dipl of Health Economics & Policy, Monash University
Naturopathic Certificate, North Coll Nat (KS Jaffery, 'Nature Cure')
Dipl Clin Nutrition, Internat Acad Nutr (R Buist, 'Orthomolecular Nutr')
Level 2 Sports Coach, (ACHPER) TAFE

Self-published by: David Joseph Vance, **Multifactor Health and Education Initiative** (MFHEI)

Self-publisher: David Joseph Vance, **Multifactor Health and Education Initiative** (MFHEI)

ISBN 978-0-9871469-3-9

Dedication:

To those sufferers, of any species, unlucky enough to bear the burden of more suffering than the average, for example cage-farmed animals, among the worst so affected, and testimony to our collective mental deficiencies that the practice is allowed to continue.

Acknowledgements:

To my mother for her excellent early education of me.

To my father for some genes, and cross-cultural experiences, and for his share of my food and lodging in early life.

To Steven L Hardie, BPsych, my thanks for gifts from his personal library, and for condescending to again eat durian with me notwithstanding his years of regression to a scavenger diet, pending possible rehabilitation on further thought, I hope.

To the University of Queensland Herston Medical Library Staff, for their competence.

To the better of the scientists whose work I have examined in my work here, for their shoulders to stand on.

To Kenneth S Jaffery, 'nature curist' (his more appalling misconceptions notwithstanding, he seems to have been correct about some things), for his shoulders to stand on.

To professor(!!...) _oger _ughes and those gormless minions of his that colluded to bring about my exclusion from the master of nutrition and dietetics course at _riffith U, not relishing the prospect of having to compete with me, nor of having me go on to point out the many defects doubtless present or to be present in any work of your own, for making me look so good in comparison – none of you will ever produce a work of science to match this; your paltry intellects will be honed in the fields of networking to secure each other's job and paycheck, clabbering together like rancid scum.

To the chaotic and causally predetermined universe for my greater share than average of good luck, the disparity being unfair nevertheless; a disparity that I will try to reduce nonetheless the predetermination, because that will make me feel like my life has 'meaning', because if I do not try, then certainly my life will have little if any 'meaning', and accordingly I will ultimately feel less satisfied than if I do try.

Contents:

Notes:

This review is written for a lay audience of average mental ability, and so some things are spelled out or repeated in different places to a greater extent than would be the case if it were written for a purely academic audience.

The review goes through Stanford's and Wrangham's books mostly in their own order, from front to back of them.

I do not adhere to a standard paragraph structure, most often starting a section of Stanford's or Wrangham's words in one line, and my comment on them immediately below on a new line – this is intended to provide a balance between providing my comment immediately next to the relevant portion of their words, and providing the opportunity to read their words in a connected manner by more readily bypassing my comments.
Also, as just here, I will often put a new point on a new line, when there is no great essential continuity of sequence between the subject matter of sentences in a block of text – it is foremost a work of science, and science is not always best written in the way usually prescribed for prose and narrative things.

In the writing format of this review, I write in the first person (i.e. "It seems to me that …" rather than "It seems that …", as I think that this is the most accurate representation of the reality – the writer (some person, myself) is expressing their thoughts and opinions on some matter, rather than the writing being generated by an infallible scientific analyzing machine, much as though that is what I aspire to be as close as possible to…

In the Detailed Critical Reviews, Stanford's and Wrangham's words from the texts of their books, the most crucial of them in my opinion, are provided verbatim (reducing the chances of my misconstruing or misrepresenting them), identified by (CS, THA) and (RW, CF), and then enclosed in quotation marks, starting with their page number:
(CS, THA) "(p150-2) abc … xyz".
My comments are enclosed in square brackets and preceded by my initials:
(DV) [abc … xyz], or,
[(DV) abc … xyz] when my comments are more inside text of Stanford's or Wrangham's that I have reproduced.

Usually 3 dots ... means that a portion of a Stanford or Wrangham sentence has not been reproduced by me, other than where it indicates a 'pregnant pause' (implying that at least some of the (, and substantial,) significance of my words is understood or understandable by the reader without my needing to write it all out) and 6 dots ... ... means that at least 1 sentence of Stanford's or Wrangham's has not been reproduced by me.

Subject matters in Stanford's and Wrangham's texts, reproduced by me, that are more separated in subject matter have a one line space in between them, whereas those that are more connected will have no line spaces.

I do not provide the full reproduction of Stanford's or Wrangham's references, just the Author's name and the date – partly to save myself scarce time and effort, and partly because I greatly dislike the citing and referencing of works that have not been properly read (much less understood...) by the writer – when I have read only a scientific journal abstract, not the whole work, I indicate that by placing "abstract only read" next to the reference.

Full right of reply to my quite destructive analysis and resulting criticisms of the science, or lack thereof, in The Hunting Apes and Catching Fire, has been offered to Profs. Stanford and Wrangham.
All forthcoming responses have been reproduced here, and any responses subsequent to the publication of this book will be added entire and verbatim, no matter how negative to this book at the Multifactor Health website.
Stanford's response was only very brief, with no further response to my response to it.

Included here (p 257) is a comprehensive previous dialogue between Prof. Wrangham and myself on my reviewing of his Chapter 16 The Cooking Enigma in Prof. PS Ungar's (Ed) (2007) "Evolution of the Human Diet The Known, the Unknown, and the Unknowable," and a cursory reviewing of the first four chapters of Catching Fire, in my book "The Natural Human Diet? The Optimal Human Diet? The known and the unknown, the possible and the probable – an analysis of the evidence" (also downloadable free from the Multifactor Health website, but, as with this book, better purchased in print at the quite low price deliberately set low as a humanitarian initiative.)

Abstract for critical review of The Hunting Apes:

Stanford's theory is essentially that human meat-eating is not very much nutritionally-driven in evolution, being more a socially and culturally evolved device that men use, in the doing, to demonstrate martial virtue as a mating display to women, and that women use, in the assessment of, as a mate selection criteria.

He takes the hunting behavior of chimpanzees and some monkeys as the evolutionary archetype and origin of this behaviour.

While he is correct to some extent, his theory is crucially deficient due to his failure to consider:

(properly) the increasing weaponization of hominins from even the earliest stages of evolution;

(at all) both the intra- and inter-group defensive needs of hominins of all stages of evolution as they become increasingly lethally weaponized;

(at all) the changing proportions of need of hominins, as they become increasingly lethally weaponized, for defense from predators versus from other hominins;

(properly) the likely range and complements of hominin foods;

(properly) the likely evolved taste preferences of chimpanzees, considering that they do not actively set out to hunt meat, taking it only when they come on it, and even then usually not in seasons of fruit plenty;

(at all) the possible effects of interactions with human hunters and meat-eating, banana-providing primatologists on chimpanzee hunting behavior;

(at all) the likely evolved taste preferences of hominins;

(at all) the use of fire and cooking to change naturally unpalatable foods into palatable foods, and the possibility that raw meat is one of those naturally unpalatable foods;

(at all) that hunting can be a proxy display (both for mating and male-male competition/dominance) of martial virtue that reduces serious damage to combatants of the same hominin group that would likely occur through the combat alternative;

(at all) that hunting can be an outlet for aggression naturally evolved to establish dominance in a primate group and reap the mating opportunities commensurate with that, reducing serious damage to participants in, or victims of, combat or attack, respectively, between/among members of the same hominin group;

My conclusion is that indeed human meat-eating is not very much nutritionally-driven in evolution, being more a socially and culturally evolved device, but that all the above factors I have listed that Stanford does not consider are of major importance and necessary to formulation of a plausible theory of how human meat eating came to be established.

Summary Review of The Hunting Apes

Summary review of The Hunting Apes
Chapter 1 The Indelible Stamp (p 3-14)

Stanford begins with a description of one day's foraging by some chimpanzees, including the eating of colobus monkeys, by calling them (p3) "The hunters ..." despite his later noting, consistent with other writers, that chimpanzees do not set out to hunt, but rather set out to gather fruits and leaves (and nuts and seeds, which foods he never addresses in the book), and only pursue prey species that they come across incidentally, and then only sometimes, and then mostly in seasons of scarcity of the foods they actively seek. In so doing he has started his book with a misrepresentation.

Questions about the chimpanzee monkey-eating incident that need to be answered to provide the correct context to determine the probable meaning of it include:

In what season did the incident occur (Gombe Park, in which the incident occurred, has more- and less-plentiful seasons regards chimpanzee plant foods)?

How hungry were the chimpanzees (seasonal starvation)?

How often do such incidents occur (and of course relate this to season)?

What percentage of dietary intake does the eating of vertebrates (not invertebrates, such as insects) make up?

Had the chimpanzees been induced to increase their meat-eating through interactions with the Gombe research station's primatologists and support staff in which they would have been able to see and smell that meat was at least being handled, and probably able to know that it was being eaten (There was a banana feeding station at the research station, which the chimpanzees made very regular use of to get and eat bananas, as well as moving at will through the research camp, other than being physically barred from entering some buildings at some times)?

Had observation of the increasingly present human hunters' and poachers' (who look more like chimpanzees than other perhaps more natural predators like leopards do) hunting activities induced any increase in chimpanzee hunting behavior at any time up to the present (Gombe is a relatively small park with humans all around it)?

Had the eating of freely or plentifully available bananas at the banana feeding station at the research station caused the chimpanzees' diet to have become lower in protein derived from plant sources than it had been before (almost certainly it did, because bananas are lower in protein:energy ratio than the average of the chimpanzee plant food diet)?

In recounting the conflict in Anthropological circles about the book Man the Hunter (Lee and DeVore 1968), regarding defects in its male-centered sexist aspect, and the growth of feminist anthropology therefrom, he concurs with most of that criticism of the book, but states (p10-11) "… the central importance of meat acquisition and meat sharing in modern and ancient human societies is simply undeniable. Meat, not only as a nutritionally desirable food item but also as a social currency that is controlled by males and therefore is a tool for the maintenance of patriarchal systems, plays an essential part in the social systems of both traditional human and some non-human primate societies."

Unfortunately for Stanford's theorizing, and very much the point in considering what might have happened, and where, in the 6-odd million years of evolution since our lineage parted company with that of our modern ape relatives, the central importance of meat acquisition and meat sharing is not at all an established fact for times before the last 40,000-odd years, a time period which may not at all be taken as representative of the millions of years before that time period regards our ancestors' population density, global location, climate (Ice ages (the last finishing only 10,000-odd years ago) and the times between ice ages, and glacial and interglacial periods (9 degrees C maximum difference) within ice ages, availability of the use of fire for cooking (which greatly changes the taste of meat), and ancestral physiological and mental capacities.

He follows with (p11) "The main thrust of my book is this: that Man the Hunter was fatally flawed, first by its emphasis on the role of cognition in meat acquisition rather than meat sharing, and second by its unconscious ignorance of the role of females in the meat-control system.", with which I agree.

Summary review of The Hunting Apes
Chapter 2 Man the Hunter and Other Stories (p 15-51)

Although Stanford correctly notes (p17) " … the bits of evidence from primate behavior, fossils, and other sources do not always build a consensual picture. Instead they point to a variety of possible scenarios, all of which probably contain some important elements of the truth.", he unfortunately fails to note (which are potentially crucially important errors in the passage):
that the diet almost certainly included, as well as the fruit and leaves noted, nuts (cracked with natural hammer and anvil stones of suitable shape, as some chimpanzees do), tubers (root vegetables) (dug up with (sharpened) sticks), and seeds and grains (some eaten younger and softer, some broken up or ground (as were some of the tubers) with suitably shaped natural mortar and pestle stones (as some chimpanzees use, Yamakoshi 1998, citation), as a very logical and likely extension of the nut-cracking with hammer and anvil theme, and some soaked to softness); and also, that they also almost certainly defended themselves (, and as a group, note…) from not only predators but other groups of hominids similarly armed (to an extent that would make them much more potentially fatal adversaries in fights between each other than they would be if unarmed), with clubs and thrown stones used very early on in the evolutionary process of spending less time in the trees –
as I will demonstrate, these failures, not corrected anywhere later in the book, major flaws in themselves, lead to a major flaw of incompleteness and thereby incorrectness in his theory.

Stanford repeatedly exaggerates, usually using sensationalizing hyperbole like "voraciously", "crave", "relish", "love", "avidly", "highly desired", "devote hours", without actually quantifying, the amount of meat that chimpanzees eat (</<< 5% of calories, Leonard et al 2007).
Given that, as he says himself numerous times in his book here, they don't actively seek meat, and quite often don't take meat when they find it, even when placed trailside experimentally, and that other primatologists are in very consistent agreement that they

don't, this is quite incorrect, or, if I were to match hyperbole with hyperbole, sheer nonsense.

Stanford fails here to understand that, accordingly, their taste preferences have probably evolved for those things that they are able to access in an energetically efficient manner, the end result likely being that chimpanzees probably do not find (raw, note...) meat as palatable as their preferred plant foods, such that in the present, it is this taste preference that directs their behaviour, not abstract considerations by them of energetic efficiency.

Also, Stanford has failed to consider the likelihood that in general sweetness (and perhaps also fattiness) taste is strong and relatively inextinguishable in apes (though gorillas, unlike chimpanzees, do not like fatty foods) and humans because the natural sweet and fatty foods are scarcer than the more readily available green leaves and more starchy legumes – therefore the more readily available green leaves and starchy foods are eaten because they are more readily available and palatable enough because they are an evolutionary staple food, alongside the rarer and more palatable very sweet (sweet fruits) and very fatty (nuts and oily seeds) foods for which there is an obvious evolutionary selection for by way of the evolution of taste preference for them – and that this is why humans, unless making a conscious effort to control themselves, often overeat on the very sweet and fatty foods that are no longer less readily available, at the expense of foods that are blander but still quite palatable enough to a physiologically hungry human whose sense of taste has not been deranged by poorly chosen (partly due to the psychological effects of advertising) modern diets – in the diet of the large majority of our ancestors, it is very likely that meat (and it would most often have been lean, not fatty, meat...) was a rare food (see p 178-183) below on the biochemical and other evidence that fairly conclusively indicates this), but unlike taste preferences driving pursuit of sweet fruits and fatty nuts and seeds, there is no evidence of a strong taste preference for raw lean meat – and of course, the hungrier the animal, such as in seasons of shortage of preferred food, the more acceptable almost anything edible will eventually become.

Stanford's correlation of meat eating, tool use, and large relative brain size in *selected* primates is plainly *very* faulty because he has not considered first the group of primates that are most closely related to humans, the apes – among which, chimpanzees eat some (only sometimes, nutritionally speaking) substantial amount of meat, whereas gorillas, orangutans, bonobos, and gibbons do not.

His correlation of brain size and cooperative hunting in nonprimate animals is likewise faulty due to selection of only those examples that are consistent with his theorizing – examples that are not consistent with it are manifold and likewise relatively obvious, for example, piranhas, ants, elephants, sharks and blue whales.

It is a central focus of my dealing with Stanford's work and theories in this book that he has failed, crucially and fatally regards his central theory here, to properly consider the defense needs for hominids, over time more and more lethally armed with respect to each other, from not only predators of other species, but from other groups of similarly armed hominids, and (via non-combat resolution of conflicts, and non-combat exercise of dominance and competitive urges) from members of their own group. Part of this fatal failure is that he does not properly address weaponry – almost certainly clubs and stones as weapons were necessary for even, and more to the point particularly, the earliest hominids.

He is correct in noting that "Our deep preconceptions influence and constrain the ways in which we theorize about the early nature of humanity." And here I note the possibility that people not wanting to change their (cooked, almost invariably, which is a very important point regards palatability and food choice and the evolution of these) meat-based diets of life-long, enculturated (psychosocially programmed) habit might seek unconsciously to psychologically rationalize that habit rather than to consider it objectively and make any changes that more properly logical considerations would indicate would best be made. And anyone who has successfully incorporated, into their ability to think about human behaviour and thinking, the most important basics of first-year university psychology, or even of psychology in the popular media, knows that psychological rationalization of one's behaviour is an outstandingly prevalent, virtually ubiquitous, thing…

Was/is raw meat, in varying stages of decomposition (rotting), palatable enough to our various ancestors that they would seek it in preference to raw fruit, vegetables, legumes, nuts, seeds, grains, and tubers (root vegetables)?

How much protein do we/they need?
Only 20g of purified amino acids in the right proportions is needed to keep an average adult male human in nitrogen balance.
And consider that human breast milk is a mere fraction of almost all other animal milks in protein:energy ratio: 1.5 vs i.e. 3.7 - 5.8g/100kcal (Oftedal 1981).
Clearly, from just these two facts, we are by evolution a naturally very low-protein diet consuming animal.

Is meat a higher quality food with respect to all nutrients required? (it isn't, because it is deficient in many nutrients - see any standard food nutrient tables… and see also my (2011) book "The Natural Human Diet? The Optimal Human Diet? The known and the unknown, the possible and the probable – an analysis of the evidence," also freely available at the Multifactor Health website, but surely better purchased in print…)

He notes that in 1966 at an anthropological conference examining the status of the world's hunter-gatherers, perhaps the foremost scientific conclusion that came out of the meeting was that the importance of meat in the diets of foraging people had been exaggerated. And that "this was deeply ironic, since the most influential and ultimately notorious perspective to emerge from the meeting came to be known as "Man the Hunter"" via Washburn and Lancaster's paper "The Evolution of Hunting", which attempted to explain that the human brain's 3.5-fold increase in size and complexity since the dawn of humanity, and the nature of human intellect, interests, emotions, and basic social life, were all the evolutionary products of the success of the hunting adaptation of our more recent human ancestors in the Pleistocene.

Stanford notes that anthropologists Zihlmann and Tanner (1976) pointed out that in some of the traditional societies that are most vaunted for the man's role in hunting, up to 85% of the animal protein obtained by a household came not from men at all, but from the less glamorous role of women gathering foods such as nuts, tubers and small animals.

This implies that (probably much) more than 85% of the total protein was obtained by the women.

Stanford failed to note the perhaps fairly obvious possibility that the human male hunting is partly an arrangement for male excitement, safer weaponized competition, and release from drudgery for themselves.

Stanford notes that "In the reaction to Man the Hunter, the fact was lost that while meat may not be the *valuable* food resource it had been assumed to be, it is nevertheless the most *valued* food resource in most human groups, including among foraging people."

This seems to me correct, and to be a crucial point.

Stanford's dealing with stone tools addresses no possibilities for their use outside of butchering carcasses (though he notes correctly that the implements of gathering made and used by women might not remain archeologically).

He fails to note that the tools referred to are by no means known to be used only or even primarily for dealing with carcasses; that the dealing with carcasses is by no means known to be more for the acquisition of meat to eat than for the acquisition of hides and sinews for making clothing and bags from; and that natural stone tools such as hammer and anvil stones used in nut cracking, and mortar (shallower depression) and pestle stones used in breaking and grinding of seeds and grains, and any rock depressions used for soaking seeds are not visible in the fossil record.

And his statement that our stone tool-using capabilities evolved at the late date of 2.5 million years ago is very probably very wrong, given that modern chimpanzees are known to use naturally formed hammer (and anvil) stones to crack open nuts, and that tool use has been well noted in many other animals (just google it…).

Stanford's treatment of the evolutionary drives for bipedalism, in rebutting various theories of contributory factors, are very faulty, in particular due to his failure to properly consider the need for defense of any and all evolutionary stage of hominid or human.

And his implication that as bipedalism allowed better access to prey and carcasses it would automatically lead to a greater and proportionately large consumption of these supposedly higher quality foods, fails badly to consider that if meat was in fact preferred less than plant foods, due to evolved taste preferences for fruits, young green leaf vegetables, legumes, nuts, seeds, and tubers (root vegetables), then meat consumption would be likely to increase only in the absence or shortage of these, such as seasonally or due to population pressure.
And, psychologically and culturally, the probably quite unpleasant taste of decomposed meat eaten (raw, not cooked and flavoured…) in situations of great hunger, might have actually, through association, caused an even greater distaste for raw fresh meat in comparison with the preferred plant foods, such that it was mostly avoided rather than sought, other than in times of great hunger, as a fallback food.

Stanford, in stating that "… in the course of primate evolution, the size of the brain has coevolved in inverse relation to the size of the digestive tract.", fails to differentiate the large from the small intestine – it is in fact said that humans have more small intestine, and less large intestine, than other apes, in the direction of true carnivores.
And then, in theorizing that "This would partially explain why fruit-eating and meat-eating primates have shorter guts; these foods are more easily digested, allowing more energy to be devoted to brain size.", he fails to consider the role of legumes, nuts, seeds, and starchy root vegetables (tubers), enough of which are macronutrient-dense foods requiring less digestion than leaves to provide energy, and that an increasing dietary proportion of any of these could very plausibly bring the same alterations in gut proportions as that brought by increasing dietary proportions of fruit or meat.

Summary review of The Hunting Apes
Chapter 3 Ape Nature (p 52-102)

Stanford notes that "Chimpanzees are alone among the four species of great apes in being frequent predators. We have no reason to believe that the digestive systems of the other apes could not process meat protein. Yet it is not consumed by gorillas at all, and by orangutans and bonobos only rarely."
Furthermore, that there is a great diversity of chimpanzee behavior in different regions of Africa regards various styles of tool use, hunting, and social behavior, and that these appear to be learned traditions similar to those passed on in traditional human societies. And that among higher primates, in which there are cultural traditions and a strong learned component to being a successful hunter, prey may not even be recognized as food items unless the hunters have been taught to do so.

Stanford notes that the majority of hunting by chimpanzees is by males.
Also, that within a chimpanzee community, males are dominant to females and at times behave quite brutally in their subordination of them. However, females actively choose their mates by inciting male competition when they are sexually receptive (advertised by the presence of fluid-filled swellings on their genital regions), and with their infants form the nuclear units of the social system. The females forage independently from males to minimize competition for food and socialize with the males mainly when it suits their own reproductive interests.
He notes that at Gombe National Park, male chimpanzees who have meat will sometimes offer bits of meat to swollen females, in exchange for which they receive matings.

He notes that Manson and Wrangham (1991) consider this issue in comparison to the kinds of intergroup relations that most traditional human societies have. They considered females, whether human or ape, to be valuable alienable resources for which males should and do compete intensely. This competition places male cooperation at a premium, since each male benefits by being part of a strong alliance. They conclude that this set of benefits leads to males (and men) tending to remain in their natal area and obtaining females from other neighboring groups. Since males who remain in the home group tend to be related to one another, the incentive to compete rather than cooperate may be lessened.

As Manson and Wrangham's paper was titled Intergroup aggression in chimpanzees and humans, it seems doubly unfortunate that Stanford does not here, nor elsewhere in the book, address defense issues.

Summary review of The Hunting Apes
Chapter 4 The View from the Pliocene (p 103-135)

Stanford states "In the case of early humans, we can be sure that their diet consisted mainly of plant foods, for which they foraged all day long just as modern apes do. We can be confident of this because our ancestors lacked any anatomical specialization for catching large prey, such as carnivorelike canine teeth or powerful claws."

He fails to mention other reasons (see also "The Natural Human Diet? The Optimal Human Diet? The known and the unknown, the possible and the probable – an analysis of the evidence," also freely available at the Multifactor Health website, though of course better purchased in print…) that might indicate that this also applies to the very large majority of our ancestors, with the exceptions of some groups in cold climates and relatively very recently as population expansion caused food pressure (for example around the end of the last ice age from 40,000(?) - 10,000 years ago).

He continues "In human ancestors, cleverness and cooperation may have evolved to a point at which planned cooperative hunting may have replaced stealth and sheer attacking power. Sociality, better weapon manufacture, and larger brain size may thus be traits that arose in response to the need to find and capture prey."

He fails to properly consider the alternative, that as a primate, for example the solitary orangutan, needs a large brain size to find and keep in mind and plan expeditions among locations of fruit (and legumes and young green leaves) in season in its range, then to that feeding complexity adding finding nuts and cracking them, finding tubers and digging them up and grinding them, finding seeds and grinding them and/or soaking them, and defending oneself from predation and other rival hominid groups, and managing intra-group social relationships, might very plausibly be at least as likely to be drivers of brain size development, as hunting.

He notes that ever since Raymond Dart's discovery and description of the Taung child in South Africa in the 1920s anthropologists have been obsessed with human carnivory, despite the fact that

none of the anatomy in early human skeletons reveals any adaptations for being predatory or carnivorous.
He failed to note that where humans did actually eat animals and collections of bones formed accordingly, the collections of those bones bias the finding of hominid/human ('hominin') remains towards discovery of carnivory rather than of plant food eating because the (lack of) plant food remains does not trigger the search for and discovery of the associated hominid remains in the same way, also that there is an almost universal failure to do what very easy calculations the data allows at such archeological sites to approximately quantify the amount of meat eaten per hominin per time (even assuming that the purpose of the butchery always included the acquisition of meat and not only hides and sinew for clothing materials and bags), which leave a greatly inflated impression of the proportion of the diet that meat eating actually comprised, both at the particular site, and even more so for hominins of the time period in general.

Stanford notes that carnivores reduce the freshly killed body of a large ungulate to a skeleton via a consumption sequence in most cases in which the consumer rapidly devoured the hindquarter flesh first, followed by the rib cage and foreleg meat, followed by the marrow within the bones, and finally it ate the contents of the head of the prey. Whereas when chimpanzees make a kill, they typically extract the brain first, which is the single greatest concentration of fat in the body, followed by the marrow of the long bones, another rich source of fat.

He states that there is some evidence to suggest that early hominids ate meat mainly in drought months in East Africa, but that eating too much of it would have been impossible, referring to the metabolic limitation for humans of no more than 50% of calories to come from protein, suggesting that carcasses, whether hunted or scavenged, would have been eaten for their fat and calories rather than protein.
He notes that another source of generally more available animal fat and protein that modern apes and human foragers eat whenever they can obtain them is insects and insect larvae. Consideration of

the seasonality of insect availability would have been a significant and appropriate improvement of his considerations here.

Stanford notes that chimpanzees scavenge very rarely, and cultural traditions may have accounted for early hominid meat-eating patterns just as they play a role in chimpanzee meat eating.
Using facts known about the hunting ecology of wild chimpanzees, he reconstructs the likely similarities and differences with the predatory patterns of the earliest hominids. The similarities were as follows:

1. The diet was primarily plant foods, with meat contributing a relatively small percentage to the overall diet. Prey, whether captured alive or encountered dead, were simply located opportunistically during the daily course of foraging for plant foods.

2. Meat was probably eaten seasonally when other foods were not available … .

3. The main influence on travel patterns was the vegetation landscape that created a mosaic of plant food foraging opportunities, not the possible but uncertain locations of meat sources.

4. Hunting was done mainly by males working together. The odds of a successful hunt were directly related to the number of hunters involved in the pursuit.

5. After a kill was made, the prize was shared with some but not all members of the hunting party, and also with females that were present.

6. The bone marrow and brain were the favored parts of the prey.

6. The possession and control of meat were used by males for their own selfish political and reproductive gains.

7. Prey were mostly small to medium sized – under 40 kg (90 lb).

The differences were as follows:

1. Stone tools used for butchering prey carcasses came into use about 2.5 million years ago.
He fails to consider their other uses, including on plant foods, and that natural stones had probably been in use, particularly on plant foods, for millions of years before that.

2. The size of the prey increased dramatically sometime after stone tools entered the fossil record, and with larger prey came the greater importance of meat in the diet.

Stanford fails to consider the possibility that not until a very long time after the advent of manufactured stone tools, in relatively recent times, was meat important in the diet, and that, as simple calculations on the data available in published work, of animals eaten per hominid/human per time period show (see also "The Natural Human Diet? The Optimal Human Diet? The known and the unknown, the possible and the probable – an analysis of the evidence," also freely available at the Multifactor Health website, though of course better purchased in print...), the importance was fairly likely to have been no more than as a very infrequent starvation or fallback food, other than very recently at the end of (the last glacial period of) the last ice age, 30,000 to 10,000 years ago, and maybe intermittently in other cold periods, but to a lesser extent due to less population pressure.

Summary review of The Hunting Apes
Chapter 5 The Hunting People (p 136-162)

Stanford notes that although early hominids possessed a different anatomy from modern people, which no doubt influenced every aspect of their behavior, and were technologically far simpler than modern foragers, and that for hundreds of years most present-day foraging societies have been in contact with other, non-foraging people, the use of modern-day hunter-gatherers (/gatherer-hunters) as examples of the range of early human social and ecological adaptations is warranted because there is no other large, highly social, tool-using biped except ourselves, and this is what *Homo habilis* was 2.5 million years ago, and these parallels alone make modern foraging people the best and only living examples of what humans do when confronted with the forces of natural selection through the rigors of the natural world.
He fails to properly consider that it isn't the natural world entirely any longer, even for them.

Also, he fails to note that modern-day hunter-gatherers (/gatherer-hunters) generally live in the more marginal areas of the world, the better (particularly for growing grain) areas having been taken over by agriculture, so that the foods available, and their proportions, and their seasons of availability, are not the same as they would be in those better areas if they had not been taken over by agriculture.

Also, when population pressure on food supplies arises, some members may emigrate to agricultural or technological populations, reducing the potential for (potentially lethal) conflict.

Furthermore, the apes are living examples of some of what our earlier (and even some of what our later) hominid ancestors would have done, and they are still very relevant to the question of what are the natural and optimal foods of humans.

Stanford asks why would human hunter-gatherers (/gatherer-hunters) search for sources of meat while chimpanzees do not, and fails to consider at all the fairly obviously plausible possibility that they do because the use of fire for cooking renders foodstuffs that might otherwise be unpalatable, for example raw meat, particularly in various stages of decomposition, palatable.

Stanford notes that chimpanzees travel quadrupedally on the forest floor (very little travel being done overhead in the trees), knuckle-walking, a mode of travel that is inefficient compared to the gait of other large-bodied animals, and that therefore chimpanzees are not well-suited for the sort of low-reward, long-distance searching that finding meat would entail. He notes that human foragers posses the bipedal posture and gait that enables them to walk slowly but efficiently for many miles, and that although the origins of our bipedal posture are not well understood, walking upright has made all the difference to many aspects of our behavior, such that when early humans became bipedal they began to search for meat actively, because only then did it become energetically feasible to do so.

However, he does not appropriately consider that if fruits, young green leaves, legumes, nuts, seeds and tubers were considered more palatable than raw meat (in varying stages of decomposition), then they may have been using the walking facility in search of these foods rather than in search of meat, and may have not eaten any meat other than when more preferred foods were not available.

Stanford cites Hawkes (1991, 1993) noting that when Amazonian Indian Ache hunters come back to camp after a successful hunt, parcels of the catch are doled out liberally and equally according to perceived need, to families, irrespective of kinship, the captor of the meat not necessarily ending up controlling its distribution, his any attempts to control it being not tolerated by the group. He notes the apparent paradox of a man going out hunting without the incentive of his success increasing what he receives. He cites Kaplan and Hill's (1985) reporting that the Ache did not receive more meat from hunters who were kin than those who were not relatives; that the percentage of food that is shared outside the hunter's family is exactly the same whether the prey was caught by one man alone or by many men working together; that the Ache hunters routinely give up so much of their catch to others that there is rarely a sharing-under-pressure scenario in the aftermath of the hunt; and that better hunters still end up giving away much more food than they can ever expect to receive.

He cites Hawkes' (1991, 1993) explanation of the apparent paradox of Ache sharing patterns. Hawkes points out that by hunting, men are targeting resources that are shared widely and are therefore available for use as politically strategic tools. Among the Ache, men show a strong preference for obtaining the foods that are the most widely shared, suggesting that they have something to gain from this acquisition, and Kaplan and Hill found that Ache women prefer the best hunters as their partners in extramarital affairs. The !Kung and the Hadza are similar, and in foraging societies, egalitarianism and an apparent lack of hierarchy are the rules that people live by.

Here, Stanford makes a fatal mistake in his thinking, by failing to consider the situation of a group that must be well bonded to be successful in fighting other groups *of the same species* in either defense or attack in lethal weaponized combat, as I elaborate on p30 below and elsewhere.

Summary review of The Hunting Apes
Chapter 6 The Ghost in the Gorilla (p 164-198)

Stanford states that it is unlikely that brain expansion in the primate order was made possible or necessary solely by predation, since only a few of the big-brained primates are hunters, but that it may, however, be necessary to be big-brained to both hunt and make use of the spoils of the hunt in a strategic way through sharing.

His focus on "Social intelligence" seems correct insofar as attributing the increase in the size of our brain to increasing natural selection pressures that favored socially and politically adept group living.

However, his discounting of "Ecological intelligence", the hypothesis for a connection between large neocortex size and the diet and pattern of food distribution among primates, on the basis of Dunbar's (1992) finding that the size (relative to the body) of the neocortex and the size of the group in which the primate lives were highly correlated across the primate order, but that there was no relationship between (relative) brain size and the home range used by a primate, suggesting that the physical environment's complexities had not led to big brains, is very probably faulty, inasmuch as predominantly folivorous monkeys are accepted to have a smaller home range as leaves are virtually everywhere, and have relatively smaller brain size, whereas predominantly frugivorous monkeys are accepted to have a larger home range as fruit is more sparsely distributed, and have relatively larger brains.

Also, fruit eaters are accepted to forage in smaller groups due to the sparseness of the distribution of the fruit, whereas folivores may congregate together while feeding as leaves are more available.

In any case, the closest primate relatives to humans, the apes, should have all been included in the studies, and it is obvious that at least the savagely solitary yet big-brained, dietarily diverse yet highly selective, orangutan was not, and this fact alone renders the Dunbar 'findings' fatally flawed regards our questions here.

Stanford notes that the roots of human intelligence may lie in a combination of ecological complexities, the value of foresight in making and using tools, and the value of being socially intelligent, and regarding the latter, he stresses frequent small deceptions, planned maneuvers of one's mates, and general politicking skills.

He notes that sharing is especially difficult to maintain in the face of those who selfishly cheat, reaping the benefits of sharing without contributing, and leave more offspring in the next generation, such that the cheating trait would spread, leading to the collapse of cooperative sharing. He states "The art of the deal among sharers is to give a bit less than you receive. Giving just enough to perpetuate the relationship with the fellow sharer is the goal."

Here, Stanford makes a fatal mistake in his thinking, of just *what* the relationship *is*, by failing to consider the situation of a group that must be well bonded to be successful in fighting other groups *of the same species* in either defense or attack in lethal weaponized combat. And, that if the animal, for example some hominid or early human, had enough cognition to remember in a detailed way, and for a long time, the lack of support in a battle won or even survived with group still viable (after having shared fully or very substantially), to the extent of resulting in the active punishment of the miscreant, to an extent that apes are not capable of thinking to do, and for example deciding to push the miscreant to the fore in the next battle, or kick them out of the group, then that sort of battle-shirking behavior would not have the same survival value that it does in monkey or ape society, but rather a negative survival value, including that females with enhanced cognition might not respond favorably with mating opportunities to a battle-shirking action witnessed. And if defection to the other group after it had witnessed the battle-shirking behavior elicited a like response, or worse (killing), then in a likewise way defection to the other group might very plausibly not be at all as feasible or compatible with breeding as it is for monkeys or apes.

And conversely, higher cognition would enable the association of one's failure to share fully with one's group fellows, with their failure to back one up fully in lethal weaponized combat, or their defection to another group agreeable to growing their size to whatever limit maximized their survival prospects regards foraging and defense by accepting a new member with no prior battle-shirking record known. This is not the situation between monkeys and apes with more limited intelligence defending against a degree of predation or attack that will be evolutionarily established as more self-limited and less potentially fatal respectively.

Furthermore, subordinate monkeys and apes are inculcated by their dominance hierarchy superiors into the habit of deferring to aggression by taking a subordinate role, by the dominance activities within the group, and therefore are not all fully bonded, to fight fully cooperatively in combat that is cognitively recognized as potentially lethal.

Also the ability to think about the future and one's survival in it as a function of one's relationships with one's group fellows would play a role that it could not in monkeys and apes.

Stanford also does not recognize, much less properly consider, that with improving weaponization and physical adaptation to weapon use, the population increases as a result of reduced vulnerability to predation (assuming cannibalization does not increase sufficiently…), and therefore supplies of more naturally preferred foods become increasingly shorter.

Also Stanford does not properly consider that the urges for competition and dominance within the group have an outlet in the hunting – the prey is the subject for domination and defeat instead of one's group-mates, and success in the hunt also serves as a proxy display of martial virtue or combat power – witnessing each other's lethal effectiveness in killing prey as large and larger than themselves is a constant reminder, for a cognitively capable hominin, that combat with each other is much more potentially fatal than it would be in the unarmed state, as is witnessing each other's lethal or potentially lethal effectiveness in defending against predators as large and larger than themselves – and this would have come in early in hominid evolution, as use of clubs very probably occurred very early on.

And there is a need for cooperation in defense against large predators, where a cognitively capable hominid in a small group (such as would be the theoretically accepted foraging party size in a mosaic environment of sparsely distributed resources, as it is in the great apes) might recognize that if his group-mates died from lack of support in defense from predation that he himself or the continually reducing group would eventually likewise succumb. Mitigating against this would be the opportunity to join other hominid groups, but maybe more so at the earlier period in time when predation not competition between hominids was the limiting factor – at a later period in time, with increased size and strength, better biomechanics for upright activity including the better use of better developed weapons, and better cognition, predation was no longer the limiting factor to survival, but competition for limited preferred and even any food, and the option of joining another hominid group was no longer so easily available, particularly if one's cheating, battle-shirking behavior had been noted and remembered by other cognitively able hominid group members, who might not see such a potential group member as a positive acquisition for the group, the opportunity to join other groups after having become known by them to have let one's original group members down in encounters of either or both defense against predators and combat with rival hominid groups might plausibly have been much less available.

These situations plausibly give rise to a selection for survival of greater sharing behavior, and reduction of cheating behavior.

In stating that "Egalitarianism is a system in which the self-interests of the individual are at times submerged in order to benefit the group. …… So the goal is not to get ahead oneself, but to ensure that no one else gets ahead.", Stanford unfortunately does not consider that the goal is very much also to ensure that no one else falls too far behind, otherwise the group will not fight potentially lethal battles with other hominin groups successfully because those who have been allowed to fall behind will not be bonded to the others in the group in such a way as to make it likely that they will fight alongside them in fully committed cooperation.

Summary review of The Hunting Apes
Chapter 7 Meat's Patriarchy (p 199-217)

Stanford, in stating that "In many traditional human societies, men hunt but women procure most of the protein and calories for their social groups through their gathering of roots, fruits, and small animals.", fails to mention legumes, nuts and grains, but correctly notes that "The value of meat is a matter of perception by group members. It is irrelevant that plant foods are as valuable a resource as meat. What matters is that animal carcasses are considered by both men and women as a more valued resource." He correctly states that "The fact that meat is so highly valued even when it composes a small part of the diet is powerful testimony to its value as social currency. Men are able to use meat to enhance status, show beneficence, and even to obtain more sex by having caught meat."

He notes that competition among males, and female choice of male mates with particular physical features, led to the 10-15 percent size difference between ancestral males and females. He cites Sanday (1981) as having surveyed a range of traditional societies and concluded that those that eat a diet in which meat plays an important part are more likely to be strongly patriarchal, and cites Adams (1990) pointing out that when males control a resource as highly valued as meat, its worth as a nutrient is largely mythical, for example because tubers and beans make an equally protein-rich diet. He notes, correctly I think, that perhaps this is because meat eating is associated with meat getting and the other masculine attributes traditionally connected to hunting, and, that meat has long been a symbol of masculinity only because it served males well throughout human evolution as a political currency that is used to enhance male alliances, snub rivals, and control females, though females have their own way of manipulating relationships such that they play a fairly equal role in deciding which genes get passed on.

His statement that "The relationship between hunting and sex is nowhere clearer than among the Sharanahua, a people of the Peruvian Amazon Basin." (citing Siskind 1973), such that, "Put at its crudest, the special hunt symbolizes an economic structure in which meat is exchanged for sex.", is flawed to the extent that in considering the representativity of the Sharanahua at the time Siskind studied them to how they lived at an earlier time, he is far too incautious in assuming that their social structure was representative of them at earlier times. In this regard note the following extract from Hern's (1991) work on Amazonial tribes in general (The bolding of text has been added):

"Generalizing from these sparse data is hazardous, but it seems that following a **catastrophic depopulation in the first 400 years following Western contact**, at least some Amazonian populations are experiencing high fertility and rapid population growth, whereas others have become extinct or nearly so. It remains to be seen whether those experiencing rapid population growth can maintain their traditional cultures in any respect.

The simple fact is that Amazon societies, however vigorous and complex, were overwhelmed by the introduction of Old World diseases and European's aggression. Black (1990) asserts that the severe impact of epidemic disease introduced from the Old World was due both to social disruption and to genetic isolation."

Accordingly, the relationships regarding defense needs at the time of Siskind's study could have been far deranged from what they once were.

Stanford, crucially, never begins to give defense needs and capabilities, crucially important in understanding the role of meat eating in human societies, their proper consideration.

Stanford's concluding attempt to downplay human natural aggression, fails to consider that often the only or major constraint on intraspecific violence is, that being of the same species, the combatants would be more evenly matched than predator and prey are, and therefore serious fighting, in the sense of attacking in the most damaging way possible, is more likely to incur, even to the victor, a degree of injury that could threaten longer-term survival. And this is more so for animals that are obligate carnivores and are therefore equipped with more potently damaging weapons like claws and fangs and body musculature to go with the use of those in a combat situation – in animals less so equipped, such as in unarmed humans, intraspecific violence is often evolutionarily determined to be greater in effort, as superiority in strength can be established with a lower risk of life-threatening injury.

And again, crucially (and somewhat obviously it seems to me), he does not address the difference between actions directed at members of another group, and actions directed at members of one's own group – evolutionarily speaking, and particularly in the case of small-group hunter-gatherers (/gatherer-hunters) it should be fairly obvious that these are likely to be two very different things. Even a basic knowledge of human history should provide ample evidence for the substantial capacity of humans to inflict violence on members of other tribal, ethnic, socioeconomic, national, religious and gender groups.
Furthermore, that the capacity for sharing with members outside our own group, or outside groups that have (more tangible or obvious) mutually-beneficial relationships already, is very obviously very limited, otherwise there would not be the great disparity in socio-economic status that there is among the nations of the world, and within the nations of the world, such that what sharing there is could often reasonably be considered very token, should be fairly obvious.

Detailed Review of The Hunting Apes

Detailed review of The Hunting Apes
Chapter 1 The Indelible Stamp (p 3-14)

(DV) [Stanford begins with a description of one day's foraging by a group of chimpanzees, including the eating of colobus monkeys, by calling them]

(CS, THA) "(p3) The hunters ..."

(DV) [despite his later noting, consistent with other writers, that chimpanzees do not set out to hunt, but rather set out to gather fruits and leaves (and legumes, and nuts, and seeds, which foods he never addresses in the book), and only pursue prey species that they come across incidentally, and then only sometimes, mostly in seasons of scarcity of the foods they actively seek.

In so doing he has started his book with a misrepresentation.

This is unfortunate because the book does have some good thinking in it, thinking superior in some ways to that of many others voicing opinions and theories, notwithstanding that I posit I will here better that thinking, though some of that bettering will be an adding to his theories, amongst that (and some points of factual detail) that I rebut.

Questions about the chimpanzee monkey-eating incident that need to be answered to provide the correct context to determine the probable meaning of it include:

In what season did the incident occur (Gombe Park, in which the incident occurred, has more- and less-plentiful seasons regards chimpanzee plant foods)?

How hungry were the chimpanzees (seasonal starvation)?

How often do such incidents occur (and of course relate this to season and the availability of foods eaten in other seasons) (and of course mention localities other than Gombe, there having been noted to be large differences in the amount of hunting between different localities) ?

What percentage of dietary intake does the eating of vertebrates (not invertebrates, such as insects) make up?

Had the chimpanzees been induced to increase their meat-eating through interactions with the Gombe research station's primatologists and support staff in which they would have been able to see and smell that meat was at least being handled, and probably able to know that it was being eaten (there was a banana feeding station at the research station, which the chimpanzees made very regular use of to get and eat bananas, as well as moving at will through the research camp, other than being physically barred from entering some buildings at some times)?

Had observation of the increasingly present human hunters' and poachers' (who look more like chimpanzees than other perhaps more natural predators like leopards do) hunting activities induced, via imitation ("monkey see, monkey do"...), any increase in chimpanzee hunting behavior at any time up to the present (Gombe is a relatively small park with humans all around it)?

Had the eating of freely or plentifully available bananas at the banana feeding station at the research station caused the chimpanzees' diet to have become lower in protein derived from plant sources than it had been before (almost certainly it did, because bananas are lower in protein:energy ratio than the average of the chimpanzee plant food diet)?]

(DV) [He goes on to pose the following questions:]
(CS, THA) "(p6) Exactly when did meat become an important part of the human diet, and how was it obtained? Were early humans savage and cunning hunters, or clever but weak scavengers? How important was meat in the diet as our ancestors' lineages evolved and diversified, and how could the eating and sharing of animal

prey have contributed to the expansion and reorganization of the human brain and cognition? What are the nutritional and social roles of meat in traditional human societies? These are questions to which anthropologists studying the fossil record have few answers."

(DV) [To which I would add, they have few good answers, but plenty of bad answers...

He notes that the fossil record is scanty regards addressing these questions, and I agree, but will elsewhere (p 236-7 below) (see also "The Natural Human Diet? The Optimal Human Diet? The known and the unknown, the possible and the probable – an analysis of the evidence," also available free at the Multifactor Health website, though surely better purchased in print...) go on to illustrate just how poorly what fossil (bones of hominins (hominids (ancestors of only humans not apes) and humans, and bones of animals with and without human stone tool cut marks) evidence there is has been used in deriving some rough quantifications of importance to these questions, in particular the question of what (mostly very small) proportion of our various ancestors' diets was of meat. He notes:]

(CS, THA) "(p8) The role of meat in the lives of early hominids has been viewed at times as crucial, at other times as minor, and at still other times as nonexistent in different eras of anthropological thought."

(DV) [In recounting the conflict in Anthropological circles about the book Man the Hunter (Lee and DeVore 1968) regarding defects in its male-centered sexist aspect, and the growth of feminist anthropology therefrom, he concurs with most of that criticism of the book, but states:]

(CS, THA) "(p10-11) ... the central importance of meat acquisition and meat sharing in modern and ancient human "

[(DV) which is not *hominid*, our more distant ancestors of different species, note]

"societies is simply undeniable. Meat, not only as a nutritionally desirable food item but also as a social currency that is controlled by males and therefore is a tool for the maintenance of patriarchal systems, plays an essential part in the social systems of both traditional human and some non-human primate societies."

(DV) [Unfortunately, and very much the point in considering what might have happened, and where, in the 6-odd million years of evolution since our lineage parted company with that of our modern ape relatives, the central importance of meat acquisition and meat sharing is not at all an established fact for times before the last 40,000-odd years, a time period which may not at all be taken as representative of the millions of years before that time period regards our ancestors' population density, global location, climate (Ice ages (the last finishing only 10,000-odd years ago) and the times between ice ages, and glacial and interglacial periods (9 degrees C maximum difference) within ice ages), availability of the use of fire for cooking (which greatly changes the taste of meat), and ancestral physiological and mental capacities.
He follows with]

(CS, THA) "(p11) The main thrust of my book is this: that Man the Hunter was fatally flawed, first by its emphasis on the role of cognition in meat acquisition rather than meat sharing, and second by its unconscious ignorance of the role of females in the meat-control system."
(DV) [To which I add: amongst ignorance about many other things, in particular those that I shall explore here.]

(DV) [A lesser mistake in passing, as I go through the book from front to back:]
(CS, THA) "(p13)… our social nature is as basic a primate trait as breathing"
(DV) [Orangutans are far from social, being very solitary – he is generally correct, but overgeneralization must be guarded against.]

Detailed review of The Hunting Apes
Chapter 2 Man the Hunter and Other Stories (p 15-51)

(DV) [Stanford provides the following typical piece of anthropological lore (with which he takes issue):]
(CS, THA) "(p15-16) Around five or six million years ago, a hairy, one-meter tall creature that looked like an upright chimpanzee left the security of the woodlands for life on the open savannahs of eastern Africa. Physically defenseless but gifted with a sharp mind, the species had carved a niche for itself by becoming bipedal, allowing the creatures to travel efficiently for long distances over open ground. Its members ate mostly fruit and leaves but also included increasing quantities of meat, both hunted and stolen from animal carcasses that they found on the savannah. Eventually they learned to modify stones into tools, which over time became effective butchering implements, and perhaps also weapons. Intelligence and sociability were their most valuable assets in coping with the risks of a dangerous world. They therefore lived in social groups or perhaps in monogamous pairs for the safety of numbers. These first hominids flourished, and their descendants became ever larger-brained and more human, eventually evolving into modern *Homo sapiens*."

(DV) [In the lead-in to his partial rebuttal of this in stating that:]
(CS, THA) "(p17) … the bits of evidence from primate behavior, fossils, and other sources do not always build a consensual picture. Instead they point to a variety of possible scenarios, all of which probably contain some important elements of the truth.",
(DV) [, he unfortunately fails to note (which are potentially crucially important errors in the passage):
that the diet almost certainly included, as well as the fruit and leaves noted, nuts (cracked with natural hammer and anvil stones of suitable shape, as some chimpanzees do), legumes (a staple food of nearly all large primates), tubers (root vegetables) (dug up with (sharpened) sticks), and seeds and grains (some eaten younger and softer, some broken up or ground (or soaked to softness) (as were some of the tubers) with suitably shaped natural stones, as a very logical and likely extension of the nut-cracking with hammer and

anvil stones theme); and also,
that they also almost certainly defended themselves (, and as a *group*, note…) from not only predators but other groups of hominids similarly armed (to an extent that would make them much more potentially fatal adversaries in fights between each other than they would be if unarmed), with clubs and thrown stones used very early on in the evolutionary process of spending less time in the trees –
as I will go on to demonstrate, these failures, which are not corrected anywhere later in the book, major flaws in themselves, lead to a major flaw of incompleteness and thereby incorrectness in his theorizing.]

(CS, THA) "(p22) When a researcher constructs a grand conceptual model of early human evolution, she or he has no choice but to resort at some point to the use of a strong analogy with some living, well-studied primate relative"
(DV) [He had also noted a little earlier in the book that chimpanzees are more closely related to humans than they are to gorillas, an assertion that seems maybe somehow wrong to me. It seems well agreed that orangutans diverged genealogically from the ancestor of all of humans, chimpanzees and gorillas, though note that both humans and orangutans have thick tooth enamel whereas the chimpanzees and gorillas have thin tooth enamel, an anomaly that seems not well addressed in the literature.]
He continues
(CS, THA) "(p22-23) There are four great apes, and any one of them could be a good analogy of the human ancestor. But of course the human ancestor was not exactly alike any of them, and perhaps was so different that to use any as exemplars may be counterproductive."
(CS, THA) "(p26) William McGrew has chastised us for thinking of "the chimpanzees" by pointing out that the degree of cultural diversity among chimpanzee populations across Africa prevents us from generalizing about the species' tool-use capabilities, hunting styles, and so on. And in the same way, paleoanthropologists err when they speak of "the australopithecine" … ."

(DV) [He repeatedly exaggerates, without actually quantifying, the amount of meat that chimpanzees eat:]
(CS, THA) "(p27) We know today that the cumulative amount of meat can be quite high because of the tendency for chimpanzees to capture several monkeys in a single hunt"
(DV) [It is far more likely that less than one monkey per chimpanzee will be captured in a single hunt];
(CS, THA) "(p30) Meanwhile, in the forests of Latin America there is another, much smaller primate [(DV) capuchin monkey(s)] that also hunts and eats meat voraciously whenever it has the chance … hunt as avidly and successfully as the chimpanzees."
(DV) [Chimpanzees in general do not "hunt and eat meat voraciously whenever they have the chance" – there is a very good consensus that they do not actually direct their food-gathering at meat, but (only sometimes) take it opportunistically if it presents itself in the search for other (and therefore it would seem more preferred) foods, and that more in seasons of a shortage of those other more preferred foods. This is actually in great contrast to the capuchin monkeys, whose foraging is (p188-90 of his own book) far more deliberately directed at vertebrates];
(CS, THA) "(p48) One of the most striking differences between the foraging behavior of chimpanzees and that of humans is that chimpanzees crave the meat of other animals but do not search for it. Instead, they forage for plant foods and eat prey animals opportunistically in the course of looking for fruits and leaves."
(DV) [That they do not in fact "crave" the meat of other animals if other foods are available is indicated just by the fact that they do not search for it, and proved by their failure to take all opportunities to get it that present themselves, particularly in the experiments of Kortlandt (1967, Kortlandt and Bresser (1963), Kortlandt and Trevor (1964), cited in Kortlandt 1984), in placing trailside live animals, carcasses and eggs, where most often the opportunity (of almost absolute ease) to eat these animal foods was passed up while the animals went on their way to seek the plant foods. And again the hardship seasonality factor.]

(CS, THA) "(p48) Certainly they [chimpanzees] are skilled hunters."
(DV) [Stanford had only just, on p25 of his book, stated that groups of chimpanzees differ widely in this, that is to say that it is very much a culturally acquired skill, not a more genetically acquired (and universally polished by imitation of the parents) one as it is in for example the cats or dogs.]

(CS, THA) "(p48-49) They could presumably obtain much larger amounts of meat if they actively searched for it. Yet there is little evidence that chimpanzees search for meat. The only factor that could reasonably account for the chimpanzees' failure to search for food is that the return rate on their energy expenditure is not enough to do so."

(DV) [Stanford fails here to understand that, accordingly, their taste preferences have probably evolved for those things that they are able to access in an energetically efficient manner, the end result likely being that chimpanzees probably do not find (raw, note…) meat as palatable as their preferred plant foods, such that in the present, it is this taste preference that directs their behaviour, not abstract considerations by them of energetic efficiency.

Also, Stanford has failed to consider the likelihood that in general sweetness (and perhaps also fattiness) taste is strong and relatively inextinguishable in apes (though gorillas, unlike chimpanzees, do not like fatty foods) and humans because the natural sweet and fatty foods are scarcer than the more readily available green leaves and more starchy legumes – therefore the more readily available green leaves and starchy foods are eaten because they are more readily available and palatable enough because they are an evolutionary staple food, alongside the rarer and more palatable very sweet (sweet fruits) and very fatty (nuts and oily seeds) foods for which there is an obvious evolutionary selection for by way of the evolution of taste preference for them – and that this is why humans, unless making a conscious effort to control themselves, often overeat on the very sweet and fatty foods that are no longer less readily available, at the expense of foods that are blander but still quite palatable enough to a physiologically hungry human whose sense of taste has not been deranged by poorly chosen (partly due to the psychological effects of advertising) modern diets – in the diet of the large majority of our ancestors, it is very likely that meat (and it would most often have been lean, not fatty, meat…) was a rare food (see p 178-183 below on the biochemical and other evidence that fairly conclusively indicates this), but unlike taste preferences driving pursuit of sweet fruits and fatty nuts and seeds, there is no evidence of a strong taste preference for raw lean meat – and of course, the hungrier the animal, such as in seasons of shortage of preferred food, the more acceptable almost anything edible will eventually become.]

(CS, THA) "(p64) Chimpanzees are the only great ape to avidly hunt and consume large quantities of meat."
(DV) [This is a misrepresentation, his own words on p66 are "Although meat is a small component of the overall diet,"].
(CS, THA) "(p68) Both females and males love meat,"
(DV) [Then why don't they actively seek it?]
(CS, THA) "(p70) Female chimpanzees are much less involved in hunts despite relishing meat."
(DV) [Ditto]
(CS, THA) "(p124) East African chimpanzees relish wild pigs."
(DV) [Ditto]
(CS, THA) "(p146) Chimpanzees, while ably equipped to kill prey and possessing a great relish to do so"
(DV) [obviously not, or they would seek prey actively – they do not refrain from seeking it actively because they understand the bioenergetics of it, but because, commensurate to its dietary importance to them, they prefer other foods, evolution having fashioned their taste preferences accordingly…]

(CS, THA) "(p202) [(DV) regarding chimpanzees] Meat is so highly desired that the whole community will devote hours to catching it, even though most female and immature community members end up with only tiny scraps. No other food commands such devotion."
(DV) [Given that, as he says himself numerous times in his book

here, they don't actively seek it, and don't always take it when they find it, and that other primatologists are in very consistent agreement that they don't, this is quite incorrect, or, if I match hyperbole with hyperbole, sheer nonsense: Leonard et al (2007) cite Stanford (1996), Teleki (1981), less than 5% of calories]

(DV) [Stanford notes (p28-29) that Shirley Strum's (1981, 1983) studies of a baboon troop (the Pumphouse Gang) in Kenya who ate meat more frequently – once per day – than any other known group of nonhuman primates, suggests (due to the catalytic behaviour of one particular adult male)]
(CS, THA) "(p29) … that aside from any energetic concerns about the rate of return in hunting by baboons, meat eating is a learned tradition that may be exterminated or initiated depending on the composition of the group. Past experience, observation, and imitation of others dictated much of the meat-eating behavior of the Pumphouse Gang. …… In the Tai forest of the Ivory Coast, the majority of adult males who did the hunting in Christophe Boesch's long-term study of chimpanzees recently died from an outbreak of ebola virus. Learned traditions of hunting may have died with them."

Stanford states
(CS, THA) "(p30-31) Meat eating, tool use, and large relative brain size therefore occur in two distantly related primate groups – apes and new world monkeys – and meat eating and related behaviors are also known in an Old World monkey, the baboon. Is this a random evolutionary convergence, or has natural selection driven the convergence of these traits? These animals are exemplars of how effective nonhuman primates can be as hunters. They are just three of two hundred primate species, so one might argue that meat eating was not a fundamental factor in the rise of the human species. The coincidence of traits among these species, however, is striking."
(DV) [It is plainly *very* faulty reasoning not to have considered first the group of primates that are most closely related to humans, the apes. This is very easily done, because there are only a small number of them, half a dozen or so – among which, chimpanzees

eat some (only sometimes, nutritionally speaking) substantial amount of meat, whereas gorillas, orangutans, bonobos, and gibbons do not.]

(DV) [Stanford goes on to consider dolphins, due to them being relatively large-brained:]
(CS, THA) "(p32) For dolphins, the food resource is fish, which are unpredictable in occurrence and widely dispersed. This patchy resource promotes foraging in small, fluid social units, paralleling chimpanzee society, which is structured around the patchy distribution of their favored fruits in African forests."
(DV) [It then requires to be explained why orangutans are savagely solitary, as they too are very fruit-focused animals. Also it remains to be explained why sharks, whose prey is also fish, and who also sometimes work as a group to hunt them, have quite small brains. And why elephants, leaf-eaters, have the largest brains on earth.
A possible factor, and well known to be one of the very most important factors in considerations of this kind, whose consideration cannot therefore reasonably be explored properly by Stanford here, is defense.
In this case, dolphins need to defend themselves against sharks, and do so by working cooperatively.
Orangutans spend nearly all of their lives in trees, and are big enough and strong enough in the arms that a mother with an infant can defend them both successfully enough on her own from the predators they face.
Gorillas require defense of infants from males outside the group, who will usually kill the infants of females that they take over (even if by default) from other males. And also, possibly, younger gorillas not yet large enough to defend themselves from terrestrial predators require defense.
Chimpanzees spend a lot of time travelling on the ground, are small enough to be vulnerable to terrestrial predators, and are well enough known to defend cooperatively.
Infant elephants are essentially totally defenseless on their own, and adult elephants cannot change direction quickly enough in the manner necessary for defense of the infant against even one adult lioness, much less a group of them.

It is a central focus of my dealing with Stanford's work and theories in this book that he has failed to properly consider the defense needs for hominids, over time more and more lethally armed with respect to each other, from not only predators of other species, but from other groups of similarly armed hominids, and (via non-combat resolution of conflicts, and non-combat exercise of dominance and competitive urges) from members of their own group. In my work here I will go on to elaborate on this aspect as being a crucial defect in Stanford's work here, in its absence.]

(CS, THA) "(p35) We see them traveling warily across the savannah, armed only with their brains against the carnivores that eagerly preyed on them."
(DV) [His objection to this vision of nascent humans or hominids as weaklings is fatally flawed regards his central theory in the book, insofar as he does not here, as elsewhere, properly address weaponry – almost certainly clubs and stones as weapons were necessary for early hominids, not only the unarmed action as a group as of chimpanzees in mobbing and driving off leopards – Early hominids were, although having some strength from daily physical activity (including daily climbing), being only a meter or so tall to start with, very likely weaker than modern humans, and weaker than modern chimpanzees (note the anecdote of a chimpanzee seizing an adult male primatologist at Gombe by the ankle and dragging him a few dozen yards through the forest as a strength display – very few modern humans of that size would be capable of such a feat, and adult chimpanzees are said to be as strong as two men, and big orangutans as strong in the arms as five men…). Also, leopards (which hunt also in trees) are much smaller than lions and saber-tooth cats/lions, and further from the trees there are pack predators, leopards being solitary hunters.]

(CS, THA) "(p37) Neither chimpanzees nor humans have any anatomical traits that specifically adapt them to a predatory way of life. Instead, both use their ability to hunt socially and cooperatively to compensate for a lack of such adaptations. One of the arguments against a hunting ancestry for early hominids has been that they lacked such adaptation and were therefore forced to scavenge for meat. This is a highly implausible scenario, in part because early hominids could make a living on the numerous small and medium-sized mammals with which they shared their habitat." (DV) [Still no mention of weaponry, even as adapted from defensive or even plant food gathering (sharp digging sticks for root vegetables) purposes.]

(CS, THA) "Our deep preconceptions influence and constrain the ways in which we theorize about the early nature of humanity. These preconceptions change, but they are always constrained by the limitations of our evidence and by the prevailing biases of the day. The extent to which these models reflect reality versus our own reflections of ourselves is embodied in some of the influential models of human origins that follow."
(DV) [Here I note the possibility that people not wanting to change their meat-based diets (cooked, almost invariably, which is a very important point regards palatability and food choice and the evolution of these) might seek unconsciously to psychologically rationalize that habit rather than to consider it objectively and make any changes that more properly logical considerations would indicate best be made. (and anyone who has successfully incorporated the most important basics of first-year university psychology, or even psychology in the popular media, knows that psychological rationalization of one's behaviour is an outstandingly prevalent, virtually ubiquitous, thing…)

Was/is raw meat, in varying stages of decomposition (rotting), palatable enough to our various ancestors that they would seek it in preference to raw fruit, vegetables, legumes, nuts, seeds, grains, tubers (root vegetables)?

How much protein do we/they need?
Only 20g of purified amino acids in the right proportions is needed to keep an average adult male in nitrogen balance.
And consider the fact that human breast milk is a mere fraction of almost all other animal breast milks in protein:energy ratio: 1.5 vs i.e. 3.7 – 5.8g/100kcal (Oftedal 1981).
Clearly, from just these two facts, we are by evolution a naturally

very low-protein diet consuming animal.

Is meat a higher quality food with respect to all nutrients required? (it isn't, because it is deficient in many nutrients - see any standard food nutrient tables… and see also my (2011) book "The Natural Human Diet? The Optimal Human Diet? The known and the unknown, the possible and the probable – an analysis of the evidence," also freely available at the Multifactor Health website, but surely better purchased in print…)

(CS, THA) "(p37-8) In 1966 about fifty anthropologists who studied the life ways of traditional foraging people gathered in Chicago for a conference to examine the status of the world's hunter-gatherers.
Perhaps the foremost scientific conclusion that came out of the meeting was that the importance of meat in the diets of foraging people had been exaggerated.

This was deeply ironic, since the most influential and ultimately notorious perspective to emerge from the meeting came to be known as "Man the Hunter." Sherwood Washburn, the most prominent and listened-to biological anthropologist of his day, and Chet Lancaster contributed a paper called "The Evolution of Hunting." It set out to explain how and why the human brain had experienced a 3.5-fold increase in size and complexity since the dawn of humanity. Washburn and Lancaster claimed that "our intellect, interests, emotions, and basic social life – all are evolutionary products of the success of the hunting adaptation" They were referring to our more recent human ancestors in the Pleistocene."

(CS, THA) "(p40) In the largest cross-cultural database that exists – a survey of 179 societies that examines how labor is divided in human groups – men alone hunt in 166, both men and women hunt in 13, and in not one do women alone do the hunting. Women, on the other hand, are the main gatherers of plant foods in about two-thirds of societies in the same survey. [(cites Murdock 1965)]".

(CS, THA) "(p40) Anthropologists Adrienne Zihlmann and Nancy Tanner [(1976)] pointed out that in some of the traditional societies that are most vaunted for the man's role in hunting, up to 85% of the animal protein obtained by a household came not from men at all, but from the less glamorous role of women gathering foods such as nuts, tubers and small animals." .
(DV) [which implies that (probably much) more than 85% of the protein of any source was obtained by the women]
(DV) [Note the possibility that the male hunting is partly an arrangement for male excitement, safer weaponized competition, and release from drudgery for themselves.]

(CS, THA) "(p41) In the reaction to Man the Hunter, the fact was lost that while meat may not be the valuable food resource it had been assumed to be, it is nevertheless the most valued food resource in most human groups, including among foraging people."
(DV) [This seems correct, and to be a crucial point, and to warrant further investigation as to the nature of that valuing, for example meat as a symbol and proxy measure of martial virtue and its application to and availability for territorial patrolling and defense; and meat as the chosen contribution of the highest of the power heirarchy.]

(CS, THA) "(p42) Stone tools made for butchering carcasses will preserve in the fossil record, while the implements of gathering made and used by women might not."
(DV) [The tools referred to are by no means known to be used only or even primarily for dealing with carcasses;
the dealing with carcasses is by no means known to be more for the acquisition of meat to eat than for the acquisition of hides and sinews for making clothing and bags from;
natural stone tools such as hammer and anvil stones used in nut cracking, and mortar (shallower depression) and pestle stones used in breaking and grinding of seeds and grains, and any rock depressions or bags used for soaking seeds are not visible in the fossil record.]

(CS, THA) "(p43) For instance, when imagining the common ancestor of all hominids, the key character is bipedalism, arising at least five million years ago and exhibited by no other primate. Our very large and complex brains, our tool-using capabilities, the increased amounts of meat in our diet, and our unusual social system all evolved at much later dates – 2.5 million years for stone tool use "
(DV) [This is incorrect, "manufacture" instead of "use" would make it correct.]

(CS, THA) " and less than 200,000 years ago for a modern level of brain-size increase. The theories of hominid origins that have gained the most attention and notoriety have been those that have woven the greatest number of human traits together in an internally consistent way, even though we will see that these often become houses of cards by virtue of the number of variables they seek to link. …… (p45) For example, here are some published explanations [(CS, THA) of reasons for the evolution of bipedalism], along with at least one piece of contradictory information:

1. Being upright gives a height advantage to intimidate predators and other hominids. Problem: Why is it important to be permanently upright? Standing upright for just a few seconds would achieve the same result.
[(DV) Stanford is not correct here, because once a predator has invested energy in approach, it is more likely to persevere to some extent with attack – some amount of perseverance is likely to be adaptive to survival.]

2. Being upright allows an early hominid to see over tall grass. Problem: Same as above, plus the doubt over whether early hominid evolution really occurred in grassland versus woodland habitats.
[(DV) Stanford is plainly incorrect here, as, particularly for the relatively slow and vulnerable animals that at least mother-with-(/and)young-child hominids are, as well as to adult male hominids that had laid down their weapons to acquire food or to feed, the

vision vigilance needs to be maintained as much as possible. The habitat was likely very often a mosaic of woodland and grassland.]

3. Being upright reduces one's exposure to intense tropical sun and heat, thereby reducing heat stress on the savannah. Problem: Again, the evidence that this key period of evolution took place on the savannah is now considered shaky.
[(DV) The evidence, such as C3/C4 plant-derived carbon (see also The Natural Human Diet? The Optimal Human Diet? The known and the unknown, the possible and the probable – an analysis of the evidence," also freely available at the Multifactor Health website, but surely better purchased in print…), seems at least mildly in favour of evolution having taken place in habitats that were substantially often at least mosaic with a grassland component.]

4. Being upright is not about walking, but rather about posture when foraging. The bipedal posture may have evolved to allow apes to pull down low-hanging, fruit-laden branches, or to allow for better tree-climbing ability on vertical trunks. Problem: Neither of these receives strong support from the behavior of modern quadrupedal chimpanzees.
[(DV) Stanford's implication that there should be strong support from the behavior of modern chimpanzees is faulty just because chimpanzees *are* quadrupedal not bipedal, and standing erect is therefore more difficult for them than it is for a bipedal animal, and because their feet having opposable toes equips them better for feeding in trees. Also this ability may be a less strong driver, that is more of an allower, of the evolution of bipedalism – and it may actually be stronger in interaction with other factors than being merely additional to them.]

5. An upright walker has its hands free for carrying food, offspring, or tools. I will deal with this last scenario below, for it incorporates some of the most widely held assumptions that have recently dogged models of human origins.
[(DV) Stanford fails to note the possibility of the improvement of weapon use – stability and mobility while swinging a club, and throwing various missiles accurately.]

(CS, THA) "(p49) …… A biped can walk longer distances in search of desired food compared to a knuckle-walker. Thus, if there was a knuckle-walking stage of hominid evolution through which early hominids passed, it would have precluded searches for unpredictable, moving sources of meat. Once bipedalism had evolved to a point of energetic efficiency, active searching could become justifiable, and meat would increase as a percentage of diet."
(DV) [If meat was preferred less than plant foods, due to evolved taste preferences for fruits, young green leaf vegetables, legumes, nuts, seeds, and tubers (root vegetables), then meat consumption would be likely to increase only in the absence or shortage of these, such as seasonally or due to population pressure. And, psychologically and culturally, decomposed meat eaten (raw, not cooked and flavoured) in situations of great hunger, might have actually, through association, caused an even greater distaste for raw fresh meat in comparison with the preferred plant foods, such that it was mostly avoided rather than sought, other than in times of great hunger.]

(CS, THA) "(p50) They suggest that in the course of primate evolution, the size of the brain has coevolved in inverse relation to the size of the digestive tract."
(DV) [Stanford fails to differentiate the large from the small intestine – it is in fact said that humans have more small intestine, and less large intestine, than other apes, in the direction of true carnivores.]
(CS, THA) "This would partially explain why fruit-eating and meat-eating primates have shorter guts; these foods are more easily digested, allowing more energy to be devoted to brain size."
(DV) [Stanford fails to mention the role of legumes, nuts, seeds, and starchy root vegetables (tubers), all of which are macronutrient-dense foods requiring less digestion than leaves (when leaf cellulose is to be digested to provide energy, which is a different requirement than that of extracting protein, minerals and vitamins from them). An increasing dietary proportion of these could very plausible bring the same alterations in gut proportions as that brought by increasing dietary proportions of fruit or meat.]

Detailed review of The Hunting Apes
Chapter 3 Ape Nature (p 52-102)

(CS, THA) "(p52-53) The hunting and meat-eating behavior of chimpanzees is a special focus, because this feature distinguishes them from the other apes and links them to the origins of our own meat-eating behavior. ...... (p54) forty years of observation in Gombe National Park, Tanzania, have only compiled information on just over three generations of chimpanzees."
(DV) [And the observation has been in both directions, chimpanzees observing humans at the research station that nearly all of them frequent and range freely through other than the inside of some buildings at some times.
Fig 3.1 (p59) indicates that the Gombe chimpanzee group's territory area has averaged about 14 square kilometers, that is, less than 4km by 4km – the significance of this is that if the research station was in the center of the area, then there would be less than 2km from the outer limit of their range to the research station and its banana feeding station, that is, the research station, and its banana feeding station, and its opportunities for chimpanzees to observe human behaviour, was within easy reach on a daily basis from anywhere else the chimpanzees might also go. The territory area was directly related to the number of males in the group, and ranged from 6 (2.5km by 2.5km) to 23 (4.5km by 4.5km) square km.]

(CS, THA) "(p59) The majority of hunting by chimpanzees is also by males. Within the community, males are dominant to females and at times behave quite brutally in their subordination of them. ...... (p60) Chimpanzees also live, however, in a society composed of females who actively choose their mates by inciting male competition when they are sexually receptive (advertised by the presence of fluid-filled swellings on their genital regions), and who with their infants form the nuclear units of the social system. These females forage independently from males to minimize competition for food and socialize with the males mainly when it suits their own reproductive interests.."

(CS, THA) "(p61) The diversity of chimpanzee behavior has become clear as studies from different regions of Africa document various styles of tool use, hunting, and social behavior. ...... Tai chimpanzees do not use fishing probes to get the termites in their forest [(DV) which the Gombe chimpanzees do], nor do Gombe chimpanzees use the abundant stones in their forest as hammers [(DV) to crack open nuts, as the Tai chimpanzees do]. These appear to be learned traditions similar to those passed on in traditional human societies."

(CS, THA) "(p66-67) Although meat is a small component of the overall diet, its role in some chimpanzee societies is disproportionately great. What chimpanzees eat when they eat "meat" is the muscle tissue, brains, bone, hair and viscera of captured animals. ...... There is little evidence that hunting is undertaken with a prior search image or intent to hunt in mind. Instead chimpanzee parties spend their days travelling the forest in search of ripening fruit and other plant foods. When they encounter prey animals they sometimes attempt to capture them. In all of the studies in which chimpanzees have been watched for years on end, the major prey they have been seen to eat is the red colobus monkey, a long-tailed treetop monkey found in various forms in forests across equatorial Africa. In addition, chimpanzees take piglets of the common wild pig species found in many African forests, fawns of bushbuck antelope, and a variety of other monkey species."

(CS, THA) "(p68) At Gombe, hunting occurs throughout the year, but is much more frequent in the dry season and during periodic "binges"."

(CS, THA) "(p68-69) In Tai national Park, the Swiss primatologist Christophe Boesch has reported extraordinary levels of cooperation among hunters. Some hunters act as drivers, pushing the colobus through the treetops toward other chimpanzees who have climbed into the monkeys' path to intercept them. Those hunters who fail to make a kill but have contributed to the overall success of the hunt end up receiving a share of the meat from the captor. ...... At

Gombe, by contrast, little overt cooperation of the Tai variety has ever been reported. Gombe chimpanzees hunt the way that many other social hunting animals do; everyone is intent on the same goal, and the sheer number of hunters trying to accomplish the same thing makes the odds that someone succeeds more likely. ...... Gombe chimpanzees capture meat and share it nepotistically – with close relatives and allies – rather than with hunting comrades."

(CS, THA) "(p69) Why should we think that cooperation is a more highly evolved art form than selfish manipulation? In our own selfish lives, each type of strategy has its purpose and rewards, and for both politicians and team athletes, different types of tasks require different talents."
(CS, THA) "(p70) At Gombe, male chimpanzees who have meat will sometimes offer bits to swollen females, in exchange for which they receive matings. These sexual manipulations show both the males ability to exert control over female reproductive behavior and possibly females' abilities to obtain meat without having to kill it themselves. Female chimpanzees are much less involved in hunts despite relishing meat [(DV) If they are hungry enough, due to scarcity of other food - seasonally juveniles often cease to grow...]. At Gombe, fewer than 10 percent of all kills have been made by females over the past twenty years. Hunting carries the risk of injury from colobus monkey teeth; hunters sometimes come away from hunts with puncture wounds on their bodies. Exposing their offspring to a colobus's canine teeth during a hunt is a risk that females with offspring should try to avoid."

(CS, THA) "(p79) This is a very important difference between the two societies: access to reproductively active females is much greater for bonobo males than it is for chimpanzee males. This may in turn account for the lower levels of aggression among male bonobos than those reported for male chimpanzees."

(CS, THA) "(p83-84) Far from the lethargic celery eaters of the mountains [mountain gorillas], these rain forest gorillas eat large quantities of fruit, and they forage far and wide to find it. ......

gorillas eat diets that are rich in fruit whenever and wherever they can."

(CS, THA) "(p94-95) Joseph Manson and Richard Wrangham consider this issue in comparison to the kinds of intergroup relations that most traditional human societies have. They considered females, whether human or ape, to be valuable alienable resources for which males should and do compete intensely. This competition places male cooperation at a premium, since each male benefits by being part of a strong alliance. They conclude that this set of benefits leads to males (and men) tending to remain in their natal area and obtaining females from other neighboring groups. Since males who remain in the home group tend to be related to one another, the incentive "
(DV) [*tendency*, due to habituation to each other's presence, may be more the point]
" to compete rather than cooperate may be lessened."

(CS, THA) "(p95) Chimpanzees are alone among the four species of great apes in being frequent predators. We have no reason to believe that the digestive systems of the other apes could not process meat protein. Yet it is not consumed by gorillas at all, and by orangutans and bonobos only rarely."

(CS, THA) "(p96) They witnessed bonobos capturing small guenon monkeys, but rather than eating them the bonobos used them as playthings. After catching a monkey, rather than apply a killing bite and consume it as a chimpanzee would, the bonobos would carry the monkeys for hours and played with them as though they were dolls. Besides monkeys, bonobo habitats in which human poaching has not been intense also contain small duiker antelope, a favorite [(DV) frequent, maybe more accurately] food for chimpanzees, plus a variety of small mammals. It does not appear that bonobos fail to hunt due to lack of potential prey."
(DV) [Note here that monkeys being somewhat similar to chimpanzees, chimpanzee predation on them could serve as an outlet by proxy for aggressive impulses to other chimpanzees in the same group, aggressive impulses due to sexual frustration not being

present in the bonobos being part of the reason for their lesser aggression to group-mates.]

(CS, THA) "(p97) This [slowness of movement] may be why orangutan meat eating is limited to the occasional squirrel found in tree cavities while foraging for fruit."

(CS, THA) "(p97) My work at Gombe has shown that the energetic balance involved in hunting rarely tips in favor of a nutritional motive. Most members of the hunting party receive very little meat for their effort, and the number of chimp-hours expended on the hunt plus the long begging and sharing session that follows it can be enormously costly relative to the quantity of meat that is usually available. The most typical catch is a one-kilogram baby, divided among up to twenty hunters. So chimpanzees engage in an energy expensive behaviour, and most fail to recoup their caloric investment. Granted, the nutrient value of meat is different from its pure caloric value; perhaps the value of some saturated fat and animal protein outweighs the relatively few calories that lean monkey meat provides. ...... (p98) But what if the motivation to hunt or not hunt is not determined by the physical environment, but by the social one? Social factors are clearly implicated in chimpanzee hunting behavior. ...... (p98-99) Communal or cooperative hunting is essential to high [hunting] success rates among chimpanzees, and chimpanzee societies vary in the level of cooperation that is seen. Among bonobos, fewer than a dozen hunts have been observed in two decades of research. Male bonobos do not cooperate or form coalitions to the same degree as do the Gombe chimpanzees. Moreover, female bonobos are given power by males in all feeding contexts due to the strategic deference by males."

(CS, THA) "(p99) In fact, chimpanzees can digest meat as well as the human body can, although meat seems to be an acquired taste not possessed by all chimpanzee societies, nor by all chimpanzees reared in captivity ...... (p100) Among higher primates, in which there are cultural traditions and a strong learned component to being a successful hunter, prey may not even be recognized as food items unless the hunters have been taught to do so. …… (p101) We see examples of this in the failure of Tai chimpanzees to attack pigs when they are encountered, or when Gombe chimpanzees show little interest in fresh carcasses of antelope they encounter in the forest. Presumably, once the innovation of meat eating has taken place, it will spread among the members of a chimpanzee community, and into other communities with the departure of emigrating females. But cultures live and die, and it is equally likely that a learned desire for meat could be extinguished only to reemerge later. In this way some populations might hunt only particular species, or even not hunt at all, due only to the halting process of the birth and death of cultural traditions."

Detailed review of The Hunting Apes
Chapter 4 The View from the Pliocene (p 103-135)

(CS, THA) "(p103) In the case of early humans, we can be sure that their diet consisted mainly of plant foods, for which they foraged all day long just as modern apes do. We can be confident of this because our ancestors lacked any anatomical specialization for catching large prey, such as carnivorelike canine teeth or powerful claws."
(DV) [This is the wrong reason for having any such confidence – our earliest ancestors of 6 million years ago had elongated canine teeth much as the apes do, and if, by the time they had became more properly early hominids, and lost the elongated canines, they had developed a culture based strongly on hunting (And Stanford has only just noted here, correctly, the cultural bases for hunting and meat eating in chimpanzees) their diet(s) might not have consisted mainly of plant foods, as for example the South American native Indians, the Ache's diet does not – I elsewhere (see also The Natural Human Diet? The Optimal Human Diet? The known and the unknown, the possible and the Probable – and analysis of the evidence, also available free at the Multifactor Health website, but surely better purchased in print…) elaborate better reasons to be fairly sure that the very large majority of our ancestors, with the exceptions of some groups in cold climates and relatively very recently as population expansion caused food pressure (for example around the end of the last ice age from 40,000(?) - 10,000 years ago), ate a very predominantly plant food based diet.]

(CS, THA) "(p104) A few archeologists, such as Jeanne Sept of Indiana University, have tried to reconstruct the likely vegetation landscape that early hominids would have relied upon, but since eating most plant foods does not require any tool use (one exception might be digging for subterranean tubers), "
(DV) [other obvious exceptions, that are potentially crucial to note, are cracking nuts with naturally formed hammer and anvil stones, and grinding seeds (and tubers) with naturally formed shallow mortar and pestle stones, and using bags to soak seeds in.]
(CS, THA) " little evidence exists of those habits.

The primary evidence we have in the fossil record for the advent of meat eating is the appearance of crude [manufactured] stone tools, beginning in East Africa in the mid-Pliocene period about 2.5 million years ago."
(DV) [Firstly, meat eating could have been done in the manner of the chimpanzees when our ancestral hominids still had the elongated canine teeth; secondly the tools could have been used on plant foods; thirdly, what use of tools there was on carcasses is just as likely to have been primarily directed at acquiring hides and sinews to make clothes, bags and other tools, as at acquiring meat to eat.]

(CS, THA) "(p105) In human ancestors, cleverness and cooperation may have evolved to a point at which planned cooperative hunting may have replaced stealth and sheer attacking power. Sociality, better weapon manufacture, and larger brain size may thus be traits that arose in response to the need to find and capture prey.
(DV) [Alternatively, if an orangutan needs a large brain size to find and keep in mind and plan expeditions among locations of fruit (and legumes and young green leaves) in season in its range, then would not adding to that, finding nuts and cracking them, finding tubers and digging them up and grinding them, finding seeds and grinding them and/or soaking them, and defending oneself from predation and other rival hominid groups, and managing intra-group social relationships, be just as likely to be drivers of brain size development as hunting?]

(CS, THA) "(p106-8) The debate about the early human diet has a long and tumultuous history. The earliest hominids almost certainly ate most of the same foods as modern apes: fruit, leaves, seeds, insects, plus some animal prey. But only the animal prey has attracted the persistent attention of anthropologists …… Raymond Dart, whose discovery and description of the Taung child in South Africa in the 1920s was the earliest evidence of a primitive hominid, believed that the first humans [australopithecines] were small but brutal killers …… According to Dart the australopithecines were "confirmed killers … that seized living quarries by violence, battered them to death, tore apart their broken

bodies, dismembered them limb from limb, slaking their ravenous thirst with the hot blood of victims and greedily devouring livid writhing flesh." ...... Anthropologists ever since have been obsessed with human carnivory, despite the fact that none of the anatomy in early human skeletons reveals any adaptations for being predatory or carnivorous."
(DV) [Note that the australopithecines did not have elongated canine teeth, so if there was elongation in a common ancestor, it was lost a long time before the use of manufactured stone tools started about 2.5myr ago.]

(CS, THA) "(p108) South African anthropologist C.K.Brain, in a landmark study, showed that the damage on australopithecine skeletons thought to have been made by other hominids was in fact caused by predators such as leopards that were feeding on these creatures. The earliest hominids were thus more likely prey than mighty predator."
(DV) [obviously enough, to anyone with even a basic understanding of combat realities, one might have thought....]

(CS, THA) "(p109) Accumulations of animal bones and human remains that had long been seen as the product of hunters and their prey were reinterpreted as natural depositions due to stream currents and geological forces that threw a misleading puzzle together for investigators eons later."
(DV) [To make this more explicit and exact regards general relevance, where humans did actually eat animals and collections of bones formed accordingly, the collections of those bones bias the finding of hominid/human (hominin) remains towards discovery of carnivory rather than of plant food eating because the (lack of) plant food remains does not trigger the search for and discovery of the associated hominid remains in the same way – there is therefore a probably great archeological bias to discovering carnivory over discovering plant food eating.
And also the almost universal failure to quantify the amount of meat eaten per hominin per time in the manner that I do elsewhere (see also "The Natural Human Diet? The Optimal Human Diet? The known and the unknown, the possible and the Probable – an

analysis of the evidence," also available free at the Multifactor Health website, but surely better purchased in print...) (even assuming that the purpose of the butchery always included the acquisition of meat and not only hides and sinew for clothing materials and bags) leaves a greatly inflated impression of the proportion of the diet that meat eating actually comprised.]

(DV) [On p110 Stanford deliberately misrepresents the ratio of carnivore tooth marks overlying stone tool cut marks to stone tool cut marks overlying carnivore tooth marks on bones by failing to give the actual ratio, which was 8:5 bones and not statistically significant, though suggestive of more scavenging than hunting having taken place.]

(CS, THA) "(p112-3) Blumenschine studied the process by which carnivores reduce the freshly killed body of a large ungulate to a skeleton. A consumption sequence was followed in most cases in which the consumer rapidly devoured the hindquarter flesh first, followed by the rib cage and foreleg meat, followed by the marrow within the bones, and finally it ate the contents of the head of the prey."

(DV) [On p114 Stanford neglects to consider than early humans engaged in scavenging carcasses would have been easy prey themselves unless well-armed and competent in weapon use, and operating cooperatively in a group – a failure fatal to his theory in this book, as I detail later here]

(CS, THA) "(p119) There is some evidence to suggest that early hominids ate meat mainly in drought months in East Africa [(cites Speth and Davis 1976)]. But eating too much of it would have been impossible [(DV) referring to the (modern human) metabolic limitation of no more than 50% of calories to come from protein]. This, according to Speth [(DV) and Davis?], suggests that carcasses, whether hunted or scavenged, would have been eaten for their fat and calories rather than protein. When chimpanzees make a kill, they typically extract the brain first, which is the single best [(DV) "greatest", which is not the same thing as "best", as many seem not to understand…] concentration of fat in the body. This is followed by the marrow of the long bones, another rich source of fat."

(CS, THA) "(p119) All this attention we are paying to the meat of large animals ignores, however, another source of animal fat and protein that is much less glamorous but perhaps more available. Modern apes and human foragers eat insects and insect larvae whenever they can obtain them."
(DV) [And would insects, as small short-lived animals, mostly develop in moist seasons and be mostly present as only the microscopic eggs during or by the end of the dry season, adding to the shortage of fat and protein present along with the shortage of carbohydrate and other nutrients at or by the end of the dry season?]

(CS, THA) "(p121) When chimpanzees kill an animal, they typically consume meat, bones, viscera, and hair, leaving no archeological trace on the ground when they are done. This is particularly true when the victim is a juvenile animal; it is gobbled up completely within a few minutes."

(CS, THA) "(p121) Chimpanzees scavenge very rarely; in the thousands of hours of observation at Gombe, scavenging of carcasses not killed by chimpanzees has been witnessed fewer than 20 times."

(CS, THA) "(p124) Whether early hominids ate carcasses or actively pursued live animals, or both, it is unlikely that they set out on their daily travels with the intention of doing so. Furthermore, cultural traditions may have accounted for early hominid meat-eating patterns just as they play a role in chimpanzee meat eating."
……
(DV) [He goes on (p125) to elaborate the possible mental requirements of scavenging; watching vultures; knowing carnivore hunting sites and times; knowing birth seasons of ungulates; and knowing migratory routes.]

(CS, THA) "(p126-8) Using what we know about the hunting ecology of wild chimpanzees, we can reconstruct the likely similarities and differences with the predatory patterns of the earliest hominids. The similarities were as follows:

1. The diet was primarily plant foods, with meat contributing a relatively small percentage to the overall diet. Prey, whether captured alive or encountered dead, were simply located opportunistically during the daily course of foraging for plant foods.

2. Meat was probably eaten seasonally when other foods were not available, and perhaps when prey birth seasons provided a bounty of fawns or infants. In the leanest season, the portion of meat in the diet may have been substantially greater than the yearly average.

3. The hunting range …… The main influence on travel patterns was the vegetation landscape that created a mosaic of plant food foraging opportunities, not the possible but uncertain locations of meat sources.

4. Hunting was done mainly by males working together. The odds of a successful hunt were directly related to the number of hunters involved in the pursuit.

5. After a kill was made, the prize was shared with some but not all members of the hunting party, and also with females that were present.
[(DV) Why assume that it followed the Gombe pattern in this way, instead of the more fully cooperative and sharing pattern of the Tai chimpanzees, particularly as he's just said the males were "working together", which implies cooperation, not mere coincidence?]

6. The bone marrow and the brain were the favored parts of the prey.

7. The possession and control of meat were used by males for their own selfish political and reproductive gains.

8. Prey were mostly small to medium sized – under 40 kilos (90 pounds). Nevertheless, the total amount of prey killed and scavenged each year was great (more than 220 pounds per adult male in the social group per year, and smaller amounts for females and adolescents). Early hominids may have been major predators on small and medium-sized mammalian species in their habitat, especially if their hunting peak was concentrated in the birth season of the prey.
[(DV) They might just as likely not have been, if their taste preferences were rather for nuts they'd cracked open with hammer and anvil stones, seeds they'd ground with shallow mortar and pestle stones or soaked, or tubers they'd dug up with sharp(ened) sticks and then ground with shallow mortar and pestle stones, than for raw meat… note that the lost elongation of the canine teeth was not regained at any time…]

(CS, THA) (p128) There are however, at least two profound differences between the meat-eating behavior of modern chimpanzees and that of emerging humans:

1. [(DV) Manufactured!] Stone tools used for butchering prey carcasses [(DV) and for other uses, including on plant foods] came into use about 2.5 million years ago.
[(DV) And natural stones had probably been in use, particularly on plant foods, for millions of years before that]

2. The size of the prey increased dramatically sometime after [(DV) probably a very long time after, in relatively recent times] stone tools entered the fossil record, and with larger prey came the greater importance of meat in the diet.
[(DV) As my calculations of animals eaten per hominid/human (see also "The Natural Human Diet? The Optimal Human Diet? The known and the unknown, the possible and the Probable – an analysis of the evidence" (p 91-92), also available free at the Multifactor Health website, but surely better purchased in print…), show (p 236 below), the importance was fairly likely to have been no more than as a very infrequent starvation or fallback food, other than very recently at the end of (the last glacial period of) the last

ice age, and maybe intermittently in other cold periods, but to a lesser extent due to less population pressure]

The list of similarities therefore applies chiefly to the earliest hominids, before the documented use of stone tools.
(DV) [They probably apply very similarly for all but the period 30,000 to 10,000 years ago, the last glacial period of the last ice age, when the climate was cold, and there was increasing population pressure in virtually all areas of the world.]

(CS, THA) (p129-30) We imagine more modern hunters to be very different – after all, they are men engaged in the pursuit of big game, armed with weapons, whereas the early forms of hunting discussed in this chapter apply to apelike creatures that use only their teeth and hands.
(DV) [This is almost certainly very wrong, and crucially fatal to his overall theory, as they almost certainly had clubs and missiles, at a minimum, if not also coarse spears.]

(CS, THA) One might expect the rise of more modern humans, with a related expansion of brain size, to be accompanied by cooperation during the hunt, greater powers of either spoken or gestural communication, and overall better hunting efficiency. *Homo erectus*, an early human species that lived in the Old World from about 1.8 million years ago to 300,000 years ago, has often been considered the first big-game hunter, capable of using fairly sophisticated weaponry to bring down animals as large as elephants. It was once an accepted, or perhaps assumed [(DV) indeed!...] fact, that humans of the early Pleistocene era were accomplished hunters. Perhaps the one archaeological site most persuasive on this account was Torralba, an ancient lakeshore on the plains of central Spain. There, Homo erectus was thought to have hunted elephants. ……
(p130-1) Speaking about the earlier interpretations of Torralba, Binford [(DV) (1987), an early taphonomist, studier of burial] states "This is a classic example of a post hoc accommodative argument: that is, the a priori assumption is invoked to give meaning to the recognized patterns." The initial interpretation about Torralba as a site of cooperatively organized human hunting

activities lasted for years, and all further investigations of the site were bound by that set of presumptions. Only later did many of the interpretations become revised: bones that were thought to have been modified as tools were actually modified by the action of water in the lake. Tools that were apparently fashioned from elephant tusks were merely tusks exhibiting natural breakage patterns. Even the degree of association between human and elephant bones was called into question. The result was that the earlier interpretations about Torralba were overturned. The assemblage of elephant bones appeared to be the result of natural processes rather than cooperative hunting. …… It appears that the emergence of a more modern pattern of hunting big game did not arise in the Upper Pleistocene in Europe until some 110,000 years ago. …… (p132) The archeological record for recent human prehistory thus seems to show that our ancestors emerged from a past in which meat was gotten by whatever manner possible, into a recent past and present in which hunting for live adult prey has become the norm."

Detailed review of The Hunting Apes
Chapter 5 The Hunting People (p 136-162)

(CS, THA) "(p137-9) It has become fashionable for anthropologists to criticize the use of [present-day] foraging people as models of human evolution, citing all the ways in which they are not appropriate illustrators of early human behavior. If we compare modern hunter-gatherers [(DV) /gatherer-hunters] to a likely portrait of early humans living one to three million years ago, major differences exist that should serve as caveats to drawing evolutionary conclusions from modern hunter-gatherer [(DV) /gatherer-hunter] behavior. First, early hominids possessed a different anatomy from modern people, which no doubt influenced every aspect of its behavior. Earliest humans were technologically far simpler than modern foragers, and were probably closer to what a wild chimpanzee does with tools today. Although we like to think that hunter-gatherers [(DV) /gatherer-hunters] live in a primeval environment that is almost certainly untrue. For hundreds of years most foraging societies have been in contact with other, nonforaging people.

(DV) [Stanford fails to note that modern-day hunter-gatherers [/gatherer-hunters] generally live in the more marginal areas of the world, the better (particularly for growing grain) areas having been taken over by agriculture, so that the foods available, and their proportions, and their seasons of availability, are not the same as they would be in those better areas if they had not been taken over by agriculture.

Also, when population pressure on food supplies arises, some members may emigrate to agricultural or technological populations, reducing the potential for (potentially lethal) conflict.]

(CS, THA) "...... (p139) Many anthropologists therefore assume that we should not use modern-day hunter-gatherers [/gatherer-hunters] as examples of the range of early human social and ecological adaptations. This reluctance is entirely unwarranted."

(DV) [It is not entirely unwarranted, it is only somewhat unwarranted.]

(CS, THA) "(p140) There is no other large, highly social, tool-using biped except ourselves. This is what *Homo habilis* was 2.5 million years ago
(DV) [Not exactly, which is the point…].

(CS, THA) These parallels alone make modern foraging people the best and only living examples of what humans do when confronted with the forces of natural selection through the rigors of the natural world."
(DV) [It isn't the natural world entirely any longer, even for them. Nor, as I note above, would it be representative of all areas of the natural world in past days. Furthermore, the apes are living examples of some of what our earlier (and even some of what our later) hominid ancestors would have done, and they are still very relevant to the question of what are the natural and optimal foods of humans.]

(DV) [After having recounted days in which hunting occurred for a group of Gombe chimpanzees and a group of the South American Indian Ache:]
(CS, THA) "(p143-4) While the prey that provided meat for the hunters is equally relished by both humans and chimpanzees "
(DV) [Raw meat is quite likely to not be "relished", but rather to be a fall-back food eaten when the more preferred and sought after fruits, nuts, legumes and seeds are in short supply or unavailable, such as seasonally, and humans and chimpanzees are very hungry.],
(CS, THA) " only the humans actively searched for their prey. For chimpanzees, the hundreds of prey animals that are eaten each year "
(DV) [Meaningless, and misleading, other than as if given as the percent amount per chimp per time period]

(CS, THA) " are essentially a bonus of protein and fat that comes to foragers in search of plant foods. For hunter-gatherers [(DV) /gatherer-hunters], foraging trips have the explicit intention of finding both meat and other foods. I have seen very few encounters between chimpanzees and their prey in which the chimpanzees changed their travel direction or in any way indicated

an intention to find prey other than simply bump into it while looking for other foods. This seems strange given the obvious desire that chimpanzees have for meat."
(DV) [That they do not in fact have an "obvious desire" for meat if other foods are available is indicated just by the fact that they do not search for it, and proved by their failure to take all opportunities to get it that present themselves, particularly in the experiments of Kortlandt (1967), Kortlandt and Bresser (1964), and Kortlandt and Trevor (1963) (all cited Kortlandt 1984) in placing trailside live animals, carcasses and eggs, where most often the opportunity (of almost absolute ease) to eat these animal foods was passed up while the chimpanzees went on their way to seek the plant foods.
And again the hardship seasonality factor.
Stanford fails here to understand that, accordingly, their taste preferences have probably evolved for those things that they are able to access in an energetically efficient manner, the end result likely being that chimpanzees probably do not find (raw, note...) meat as palatable as plant foods, such that in the present, it is taste preference that directs the feeding behaviour – and of course, the hungrier the animal, such as in seasons of preferred food shortage, the more acceptable almost anything edible will eventually become.]

(CS, THA) "Why would human hunter-gatherers [/gatherer-hunters] search for sources of meat while chimpanzees do not?"
(DV) [Because the use of fire for cooking renders foodstuffs that would otherwise be unpalatable (such as raw meat, in various stages of decomposition), palatable, an obvious enough reason to consider, in the absence of psychological bias not to consider it, one might have thought....]

(CS, THA) "...... (p145) Chimpanzees travel quadrupedally on the forest floor; very little travel is done overhead in the trees. On the ground, chimpanzees knuckle-walk, a mode of travel that is inefficient compared to the gait of other large-bodied animals. Even though chimpanzees may walk several miles per day in the wild, they are generally travelling to known locations of fruit trees and

don't knuckle-walk aimlessly around the landscape. Chimpanzees are not well-suited for the sort of low-reward, long-distance searching that finding meat would entail. Human foragers posses the bipedal posture and gait, enabling us to walk slowly but efficiently for many miles. Many human foragers travel further on a foraging day than chimpanzee parties ever travel. The origins of our bipedal posture are not well understood, but walking upright has made all the difference to many aspects of our behavior, and meat-eating may be one of these. …… This [(DV) energetic efficiency of hominin bipedal walking] suggests that when early humans became bipedal they began to search for meat actively, because only then did it become energetically feasible to do so."
(DV) [However, if fruits, young green leaves, legumes, nuts, seeds and tubers were considered more palatable than raw meat (in varying stages of decomposition), then they may have been using the walking facility in search of these foods rather than in search of meat, and may have not eaten any meat other than when more preferred foods were not available.] ……

(CS, THA) "(p146) … most chimpanzee prey consists of small monkeys that weigh less than 5 kilos (11 pounds)." ……
(CS, THA) "Chimpanzees, while ably equipped to kill prey and possessing a great relish to do so "
(DV) [obviously not, or they would seek prey actively – they do not refrain from seeking it actively because they understand the bioenergetics of it, but because, commensurate to its dietary importance to them, they prefer other foods, evolution having fashioned their taste preferences accordingly…],
(CS, THA) " do not rely on meat like [to the extent that] even the most meat-impoverished "
(DV) ["least meat-consuming" is the appropriate term, as the question of the optimal proportion of meat consumption, nutritionally, for humans, is still just that, a question, and the answer may very well be a very low proportion indeed, and that rather of insects than meat per se]
(CS, THA) " human societies do."

(CS, THA) "(p147) Although hunter-gatherers [/gatherer-hunters] hunt alone or in groups, cooperation among the hunters [(DV) (chimpanzees)] can increase hunting success dramatically (fig 5.1)" (DV) [This is very wrong, there are only 9 data points on the graph, and clearly a line of best fit would slope down, not up, indicating the opposite, that the smaller the number of hunters, the larger the kilos of monkey meat per hunter – the caption to the figure ".... Hunters get more meat in the smallest and largest parties." Is itself clearly in opposition to Stanford's statement, but also it is quite obviously incorrect itself in that while an overall decreasing relationship might be statistically significant, the claimed U-shape would not be, due to the very small sample size (9) and the small difference among the height of the points in the right arm of the U relative to the differences among the left arm of the U – more of a reverse J than a U – this sort of distortion of the obvious facts to suit one's own theory is not good science, and it is not unique in Stanford's book here, notwithstanding that he has provided some thinking that is superior to much of what else there has been in the field.]

(CS, THA) "(p148) Cultural differences between chimpanzee populations pale in comparison to the rich variety of hunting techniques and meat-eating cultures that exist in traditional human societies."

(CS, THA) "(p151) Wild chimpanzees eat a diet that contains a percentage of meat so small that we cannot be sure that it plays any major role in their nutritional well-being. But in human foragers, it seems clear that meat is both highly desired and highly beneficial nutritionally. Why successful hunters even bother to gather, and why they would try to include certain resources in their diets is not clear."
(DV) [This shows a gross ignorance of human nutrition – just to start with regarding carbohydrate:fat:protein ratio, Vitamin C, vitamin E, and calcium (see also "The Natural Human Diet? The Optimal Human Diet? The known and the unknown, the possible and the probable – an analysis of the evidence," also available free at the Multifactor Health website, but surely better purchased in

print…)]

(CS, THA) "(p153) In order to explain why the Ache hunt, Kristen Hawkes [(1991, 1993)] has suggested that we should consider what the Ache do with meat once they have captured it. When hunters come back to camp after a successful hunt, parcels of the catch are doled out liberally to families. The captor of the meat does not necessarily end up controlling its distribution, and his attempts to control it are not tolerated by the group. …… (p154-6) Nutritionally, Kaplan and Hill [(1985)] found that the advantage of receiving shared meat from unpredictable game was greater than the cost involved in sharing from one's own kill. Paradoxically, even though some hunters contribute huge quantities of meat to this public good and others very little, there was no correlation between bringing home a great prize one day [(DV) yield on average, not on one day, is the correct variable to be looking at] and receiving an equivalent or larger share in return at other times. In other words, if this is a system of cultural reciprocal altruism to reduce the risk to everyone of going hungry, it still does not explain why big game is shared so widely. There is a paradox here: Why should a man go out hunting at all if his success does not increase the bounty he will eventually receive from other hunters? Kaplan and Hill consider four possible explanations. First, we should expect meat sharing to follow lines of kinship … However, Kaplan and Hill found that Ache did not receive more meat from hunters who were kin than those who were not relatives. Second … But the percentage of food that is shared outside the hunter's family is exactly the same whether the prey was caught by one man alone or by many men working together. Third … However, the Ache hunters routinely give up so much of their catch to others that there is rarely a sharing-under-pressure scenario in the aftermath of the hunt. Finally, Kaplan and Hill ask whether sharing could be accounted for by reciprocal altruism. …… However, the better hunters still end up giving away much more food than they can ever expect to receive. …… (p156-8) Hawkes [(1991, 1993)] takes an alternative approach to explain the paradox of Ache sharing patterns. Hawkes points out that by hunting, men are targeting resources that are shared widely and are therefore available for use as politically strategic tools. …… While females can create better reproductive opportunities for themselves primarily by bettering their own and their offspring's food intake, males depend on having access to as many females as possible for their own reproductive benefits. … Among the Ache, men show a strong preference for obtaining the foods that are the most widely shared, suggesting that they have something to gain from this acquisition. Hawkes reasons that Ache men are showing off by sharing a bonanza of meat. …… Kaplan and Hill found that Ache women prefer the best hunters as their partners in extramarital affairs. Women considered good hunters to be more desirable as sexual partners, if not necessarily as husbands. The sharing of big game meat by Ache men may therefore fall into the category of a mating tactic rather than family provisioning. …… (p159) Among the !Kung and the Hadza, even the best hunter must be modest. Humility is a strong cultural tradition in nearly all hunter-gatherer [/gatherer-hunter] societies. This is an extension of the egalitarian nature of these cultures. Attempts to use one's hunting prowess as an entrée to greater ambitions within the society are usually met with stern opposition, ridicule, and attempts to shame the self-promoter. It is preceded by much verbal taunting and badgering from the other members of the hunting party. …… There is no alpha male in hunter-gatherer society, no top-ranking man who dictates what the group will do and where it will go [(DV) never?...] …… (p161) But in foraging societies, egalitarianism and an apparent lack of hierarchy are the rules that people live by. Why do great apes and nearly all modern people live in rigidly hierarchical societies, while foragers do not appear to posses the concept of hierarchy? [(DV) other than male-female hierarchy?...] …… (p162) Erdal and Whiten suggest that at some point in human evolution, the cognitive ability to enforce a reverse dominance hierarchy arose, which in turn placed a great natural selection value on the ability to be politically astute in dealing with group mates. How this astute political intelligence emerged and how it has directed the evolution of our behavior are the subject of the next chapter."

Detailed review of The Hunting Apes
Chapter 6 The Ghost in the Gorilla (p 163-198)

(CS, THA) "(p164) It is unlikely that brain expansion in the primate order was made possible or necessary solely by predation, since only a few of the big-brained primates are hunters. It may, however, be necessary to be big-brained to both hunt and make use of the spoils of the hunt in a strategic way."

(CS, THA) "(p165) This chapter is premised on a Darwinian theory of mind that considers the psyche as organic an adaptation as any other aspect of our biological makeup, and applies that point of view to the cognitive aspects of hunting and meat-sharing in humans and in non-human primates."

(CS, THA) "(p166) In Darwin's view, early humans were freed from the use of their hands when they became upright walkers, later becoming proficient tool users. Natural selection favored those individuals who were bright enough to make better and better tools for survival skills, and the end result was a big-brained, bipedal, manually dexterous hominid. Unfortunately for Darwin, we now know that bipedalism arose with the first hominids at least five million years ago. There is no solid evidence of stone tool use before 2.5 million years ago "
(DV) [This is fairly obviously incorrect; from observations of modern chimpanzees using naturally shaped hammer and anvil stones to crack nuts open, it seems very likely that even the earliest hominids cracked open nuts with naturally shaped hammer and anvil stones, and likely also ground seeds and tubers with naturally shaped shallow mortar and pestle stones, none of which would be visible in the archaeological record],
(CS, THA) " so the transition from apes to hominids was not stone-tool driven."
(DV) [May well be incorrect - it may well have involved use of stone, wood, hide and bone tools in interaction with some other factor(s)].

(CS, THA) "And the expansion of the brain's volume occurred even later, much less than one million years ago."
(DV) [This is not correct; the brain had been expanding before this time (Bailey and Geary, 2009), just at a slower rate, and a case might even be reasonably made that the expansion followed the same exponential curve throughout up to the physiological limit, rather than having different linear phases of expansion, perhaps occurring as a plurality of factors became increasingly involved not only singly in an additive fashion, but also multiplicatively in an interactive fashion. Bailey and Geary found in their multivariate modeling that population density was the best predictor of brain size, which is quite consistent with my own theory (and surely some others have thought likewise in a general sense) that defense needs are important, and that population pressure on food resources is related to increased meat eating.
And note that most of the evolution of the human language ability may have occurred after the brain had reached its final modern size, if not its final neuronal arrangement. Also note the very late evolution, about 50,000 years ago (Lieberman, 2007) of the supra-laryngeal vocal tract, necessary for the full modern range of human speech sounds.]

(CS, THA) "(p166-7) Each November, Gombe chimpanzees feast on a seasonal bounty of termites. … As the dry season gives way to the first rains. "
(DV) [In other words, the season of greatest lack of the chimpanzees preferred plant foods] " During the season of termite foraging, chimpanzees sometimes travel from mound to mound all day long."

(CS, THA) "(p169) "Ecological intelligence" is the hypothesis that, as early hominids began to use increasingly complex environments, the brain's neocortex, or cerebrum, increased in volume and also underwent a reorganization to navigate the food resource labyrinth. In a lightly wooded African grassland mosaic, patchily distributed fruit trees might be farther apart than in the deep forest. Emerging hominids' large brains may have allowed them to navigate their way to distant food sources. There is good evidence for a connection between large neocortex size and the diet and pattern of food distribution among primates."

(CS, THA) "(p172) None of these requires a very impressive brain; fish live in schools and birds flock with unimpressive cognitive abilities."
(DV) [Ok, but evolution selects for light brains in birds, because they fly, and many birds are now thought to be much more intelligent than many people once thought due to birds having small brains (aside from the phenomenon of 'denigrating the victim' as a way of attempting to justify violence towards them…)].

(CS, THA) "But living in social groups and remembering all the debts and credits of life is very different. This requires the ability to recognize the other members of one's group as individuals, and to keep a running scorecard of the other group members' relationships to you and to one another over a period of months or years. We know that nonhuman primates routinely perform such cognitive feats. … (p173) Many animal behavior researchers attribute the increase in the size of our brain to increasing natural selection pressures that favored socially and politically adept group living. Primates that could best manipulate their social surroundings to their own advantage reap more mates and leave more offspring. … (p174) There can be no more basic goal of life for an intelligent creature than to get one's fellow group members to do what you want them to do without knowing that they are being manipulated."
(DV) [Ok, but in some cases it may be better that they are aware of what the manipulation is, such as if it benefits them, and they can themselves apply the manipulation and the education about it to others, for further benefit to all concerned – this might be a less basic goal – and one that evolution might have programmed us to derive pleasure from to a greater extent than from carrying out manipulations where the manipulated is unaware of what is happening.]

(CS, THA) "(p175-6) The brain increased incrementally in size throughout nearly all of primate evolution. Then, since the time of Homo erectus and the gradual transition into modern humans some

200,000 years ago, human brain size exploded."
(DV) ["exploded" is misleading insofar as the growth may have been on the one exponential growth rate increase, rather than suddenly growing at a much faster rate after some threshold point was reached.]
(CS, THA) "This explosion happened at a time when humans were probably living in small bands, leading a nomadic or semi-nomadic life of hunting and gathering. They were also probably evolving a qualitatively different and more advanced form of language and speech than had existed in earlier hominids. It may be that the brain size increase that took place during this time was due to the value of a larger and more sophisticated brain for survival and reproduction [(DV) How else?!...]. As the complexity of human societies grew, so did the pressure that natural selection brought to bear on the ability to be socially and politically clever. The result is a species in which frequent small deceptions, planned maneuvers of one's mates, and general politicking skills count for more in nearly all arenas of life than do physical size, strength, or agility. The roots of intelligence may lie in a combination of ecological complexities, the value of foresight in making and using tools, and the value of being socially intelligent."

(CS, THA) "(p176-7) British primatologist Robin Dunbar compared data from the life histories of a range of primate species to examine how the ecological function of intelligence stacks up against its political value. He found that the size [(DV) relative to the body] of the neocortex and the size of the group in which the primate lives were highly correlated across the primate order. ... Dunbar found no relationship between [(DV) relative] brain size of the home range used by a primate, suggesting that the physical environment's complexities had not led to big brains."
(DV) [This seems very suspect, inasmuch as predominantly folivorous monkeys are accepted to have a smaller home range as leaves are virtually everywhere, and have relatively smaller brain size, whereas predominantly frugivorous monkeys are accepted to have a larger home range as fruit is more sparsely distributed, and have relatively larger brains. Also, the fruit eaters are accepted to forage in smaller groups due to the sparseness of the distribution of

the fruit they seek, whereas the folivores may congregate together while feeding as leaves are virtually everywhere.]

(CS, THA) "Nor did the extent to which the species used "extractive foraging" (tool use) to obtain food correlate with [relative] brain size."
(DV) [This seems dubious – there were only 10 species in Fig 6.1 ("adapted from Byrne 1995") which seems to be attributed to Dunbar – and we are not told how they were selected – and perhaps tool use is less possible or necessary up in trees than it is on the ground].

(CS, THA) "Dunbar's findings suggest that while the first push towards a larger brain may have been the result of patchily distributed, high-quality diet and the cognitive mapping capabilities that accompanied it, the evolution of the very large brain of the higher primates was primarily due to its value in social intellect (fig. 6.1)."
(DV) [This still leaves the very solitary orangutan with its very large brain and high intelligence, and its diet of many plant foods favoring fruit, to explain. Also, it seems that the opportunities for political manipulation within the small sole adult male-dominated gorilla group (most of whom favor fruits) are much less than those available to chimpanzees, bonobos, and various monkeys]

(CS, THA) "There are many confounding factors here, since lower primates such as prosimians usually live in small groups while monkeys and apes reside in bigger groups."
(DV) [If it is known that there "are" many confounding factors, then they should be named, otherwise "may be" should be used instead of "are", and the one known possible example given. In any case, since relative brain size is the quantity dealt with, it still needs to be explained why it might be or is a confounder – for example, and very unlikely to be correct: because it is not possible for lower primates such as prosimians to live in large groups because they are smaller animals and small animals ……]
(CS, THA) "This may simply reflect their evolved predispositions for particular grouping patterns."

(DV) [It doesn't "simply" reflect their evolved predispositions, the point, very much because that is exactly the topic that is being dealt with, is to attempt to explain why.]
(DV) [In any case, the closest primate relatives to humans, the apes, should have all been included in the studies, and it is obvious that at least the orangutan was not, and just this fact alone renders the Dunbar findings fatally flawed with regard to the question they attempt to answer.]

(CS, THA) "...... Does the raw number of potential social relationships dictate a more complex social life? We assume so but have few pieces of empirical support for the idea."
(DV) [Considering the examples of small schooling fish, and communal insects, obviously enough, though not always, it merely allows it.]

(CS, THA) "This is important in considering how the modern human brain came to reach its current size. The modern (p178) human brain is on average three and a half times larger in volume than a chimpanzee's (1400 cm^3 compared to 400cm^3). But what evidence is there that early human societies were more complex than modern human societies? If group size is the primary factor, then how do group sizes compare between [(DV) present-day] foraging people and chimpanzees or bonobos? The evidence is weak, since we cannot assume that ancient humans lived in bands that were comparable to those of modern foragers. There is wide variation in group size among modern [(DV) (present-day)] foragers and also among chimpanzees."
(DV) [It seems likely that the food source complement of the early hominids was more complex than that of the apes, as it came to include (as well as most of the types of foods most of the ancestral apes (nevertheless more diverse than modern apes) were seeking) tubers and grass seeds;
that the tool use came to be more complex as the more basic use of naturally shaped hammer and anvil stones to crack nuts developed by diversification into the slightly more cognitively demanding grinding of seeds and tubers with naturally shaped shallow mortar and pestle stones;

and that scavenged animal flesh (along with the non-dietary use of hides, tendons and bones) was incorporated although (usually very) minimally.
And these seem to be very plausible activities for the early hominids to be engaged in. Keeping all of these food sources and related tool uses in mind, in the larger (compared to the apes and monkeys) mosaic area that came to be their home range, these could very plausibly be sufficient in themselves to require some growth in brain size (or change in function) from that of a modern ape, as life accordingly would probably be sufficiently more complex enough to require it.]

(CS, THA) "(p180-1) Whatever the key influences have been on primate brain size, the much more recent expansion of the neocortex is the most difficult to account for. The selection factor may have been language, which is without fossil evidence and is therefore in the realm of informed speculation."

(CS, THA) "(p181) [(Lionel)] Tiger and [(Robin)] Fox [(1971)] argued that modern humans employ behavioral strategies whose predispositions were molded in earlier eras of our evolutionary history. If we want to understand why we choose a particular mate or we strive to climb the corporate ladder of success, we must pay attention to our evolutionary past."

(CS, THA) "(p182) The arbitrary choice of the Pleistocene for the environment to which the mind is adapted ignores the fact [(DV) probability] that [(DV) some of] the roots of human cognition must be sought much deeper in our past, …"

(CS, THA) "(p182) A second problem with evolutionary psychology lies in the lack of rigor and the overabundance of [(DV) "]philosophy[(DV)"] that many evolutionary biologists perceive in the paradigm. … beyond [(DV) (unable to be)] hypothesis testing."

(CS, THA) "(p185-8) Whether intelligence is an important factor in the hunting tactics of chimpanzees is a key issue because of the putative importance of hunting to the evolution of human intelligence. ...... If females prefer to mate with the best hunters, perhaps because the females benefit nutritionally from gifts of meat, then hunting performance would be subject to sexual selection as well."
(DV) [But do females mate for other food gifts also, such that hunting is only a proxy for general foraging ability (and the evidence is that complex foraging does require more intelligence to do more successfully)?]

(CS, THA) "Since little evidence exists that Gombe chimpanzees set off each morning with the intention of finding meat, optimizing foraging routes to take advantage of the likely locations of colobus [monkey, prey] is probably not part of the hunting strategy. ...... Certainly predatory behavior involves little higher cognition for many animals. ...... In many mammalian predators, however, hunting tactics and the expected behavior of the prey species must be learned. ...... But even scavenging would have provided natural selection with fertile ground to enhance the spatial memory and navigational skills of early humans. Skills that might be important to making a kill, such as the flexibility to respond strategically to rapidly changing circumstances, should also be selected for and enhanced. ...... Among the higher primates, those species that regularly eat animal protein are the same species for which higher cognitive attributes have been reported."
(DV) [Considering the apes, which are the most important to consider because they are our closest relatives, this can be seen to not apply much at all, as only the chimpanzees meet the criterion, whereas the other three great apes and the Gibbons do not – and for the other primates, and in general, we have potentially the same type of questions as to what criteria was used to assess the intelligence, as we have in cross-cultural issues in assessment of human intelligence. Again, hunting ability might very well be just a proxy for foraging ability in general, and of no more importance than merely adding to the complement of things that must be dealt with.]

(CS, THA) "(p190-3) Christophe Boesch has claimed that highly cooperative hunting is a sign of advanced intelligence, because it requires coordinating the actions of two or more individuals in pursuit of a common goal. However, many predators cooperate when hunting without the benefit of a highly elaborated neocortex." [(DV) piranhas, ants, sharks]
(DV) [Hunting can be accomplished by mental or physical faculties, or a combination of the two; likewise, gathering].

(CS, THA) "It may be unfair [(DV) "incorrect", better] to consider cooperation a more advanced state of intelligence than selfish manipulation, deception, and nepotism, as much as we humans would like to think otherwise. We need to consider the behavior of meat eaters not while they are hunting but rather when they are dividing the spoils to see evidence of cognition that distinguishes big-brained primates from other animals. ...... As the social intelligence school of thought rose to the fore, researchers began to describe how individuals use their power and privilege to control others, to obtain more resources for themselves, and to be astute observers of their own neighbors. Among the behaviors that are of most interest is food sharing. This is due to the cooperation and reciprocal altruism that sharing entails. ...... From a Darwinian perspective, the sharer ought to reap more than he invests, for without that selfish impetus he would not share. The difficulty imagining the evolution of sharing among both humans and other animals is that such cooperative behavior is difficult to establish. Sharing is especially difficult to maintain in the face of those who selfishly cheat. If one individual cheats the system by reaping the benefits of sharing without contributing, that cheater will benefit. If sharing is at all genetically influenced, the cheaters will end up leaving more offspring in the next generation, and the cheating trait would spread, leading to the collapse of cooperative sharing."
(DV) [And here, Stanford makes a fatal mistake in his thinking by failing to consider the situation of a group that must be well bonded to be successful in fighting other groups *of the same species* in either defense or attack, in combat that is recognized as highly potentially lethal due to weaponization, in the context of group members remembering their treatment by other members.]

(CS, THA) "An example from my own work on antipredator defense makes this clear. Six or seven male red colobus monkeys mount a cooperative defense when attacked by chimpanzees bent on eating them. This is a risky business, since males are occasionally killed for their heroism. In the midst of a chimpanzee-colobus battle, I have seen male colobus watching the action from a safe perch a bit distant to the fray. By failing to codefend, the bystander reduces his own risk of injury while allowing his group mates to bear that risk. He is cheating the system. Why some males would stay back from the counterattack while others leap into it is unclear; perhaps those males who have reason to think they have fathered babies in the group " [(DV) Monkeys almost certainly do not think about that sort of thing!!]
" have a reproductive incentive to protect those progeny."
(DV) [What is actually happening is that the negative genetic survival value of a monkey behaving like that is outweighed by the positive genetic survival of aspects it is *a marker for*: aggression, dominance, size, and testosterone greater than his group-mates, and which make him (and his type) more successful overall in siring offspring in any case before they meet their death, maybe earlier through chimpanzee predation, or maybe later through having less social stress than the group-mates they inflict stress and injuries on through their dominance of them.]
(DV) [And, if the animal, for example some hominid or early human, had enough cognition to remember in a detailed way, and for a long time, the lack of support in a battle won or even survived with group still viable, after having shared fully or very substantially, to the extent of actively punishing the miscreant, to an extent that apes are not capable of thinking to do, and for example deciding to push the miscreant to the fore in the next battle, then that sort of battle-shirking behavior would not have the same survival value that it does in monkey or ape society, but rather a negative survival value, including that females with enhanced cognition might not respond favorably with mating opportunities to a battle-shirking action witnessed. And if defection to the other group after it had witnessed the battle-shirking behavior elicited a like response, or worse (killing), then in a likewise way defection to the other group might very plausibly not be at all as feasible or

compatible with breeding as it is for monkeys or apes.]

(CS, THA) "Males without such incentive might instead seek to protect their lives in hopes of mating more in the future."
(DV) [Monkeys almost certainly do not think like this!!].

(CS, THA) "In the pattern of sharing we see the roots of both sides of our altruistic and selfish natures. The art of the deal among sharers is to give a bit less than you receive."
(DV) [Not if the group is to be fully cohesive in a potentially fatal fight, it isn't!]

(CS, THA) "Giving just enough to perpetuate the relationship with the fellow sharer is the goal."
(DV) [The point is just *what* is the relationship, and the situation has not been posed correctly here, as the situations between monkeys with more limited intelligence defending against a degree of predation that will be self-limited is not the same as that of hominins defending against highly potentially fatal weaponised attack from other hominins that is not necessarily self-limited at all, and almost certainly mostly not to the same extent.]

(CS, THA) "This subtle form of cheating is common among sharers of many primate species, including humans. Whether it is subject to penalty or correction is largely determined by the relative status of the subject and the sharer. ...... This is the situation in which a subordinate member of the chimpanzee dominance hierarchy may find himself. He is unable to negotiate from a position of strength but needy of whatever social alliances he can get, no matter how one-sided. In such instances sharing may appear reciprocal to the casual observer when in fact the powerful are manipulating others to their advantage."
(DV) [In *this* situation he will *not* be prepared to fight to the death alongside the other if it can be avoided – and his inculcated habit of deferring to aggression by taking a subordinate role will itself have inclined him to a more limited response to meeting aggression with aggression.]

(CS, THA) "...... (p194) Christophe Boesch has argued that at Tai the enormous height of the rain forest canopy to which the colobus retreat makes cooperation in hunting essential, with the result that helping out is rewarded later. At Gombe, with a lower forest structure and greater ease of hunting, such cooperation is not necessary and so no such reward system has developed."
(DV) [Another example of why the exact situation determines the nature of the sharing relationship. And in the situation of lethally-armed hominin groups fighting each other for what is cognitively recognized as including longer-term survival that is dependent on limited food resources, overlaid on other things like competitive and dominance drives, the need for an even greater level of cooperation among members of one hominin group becomes very plausible.]

(CS, THA) "(p196) I suggest that the ability to make use of meat for nutritional purposes is facilitated in a social primate by a relatively high degree of intelligence, because of the complexities of sharing the meat of other animals. Being a sharer requires a degree of encephalization that is seen only among the great apes and humans."
(DV) [This is very possibly not correct – many animals bring food back to their young even when they are hungry themselves, and this appears to have very little requirement for intelligence – perhaps the two things that might necessitate greater intelligence over that required merely for another version of sharing becoming genetically imprinted in roughly the same way that it did for infant feeding might be that the food is not the more truly naturally preferred food of the species and therefore its consumption is not as much a genetically determined routine event; and, that the package it comes in, the carcass, is not as readily divided into pieces as the more truly naturally preferred food of the species, such as fruit, which can usually be plucked from the same source that another is plucking from, whereas shared meat must instead be ripped forcibly from the carcass by the possessor and handed over to the other.]

(CS, THA) "(p196-8) Sharing and egalitarian relationships are two hallmarks of the foraging people of the modern world. The open reciprocity that occurs routinely in these human societies is rare among nonhuman primates. There is a stark contrast between the egalitarian nature of hunter-gatherer [(DV) /gatherer-hunter] bands and the strongly hierarchical structure of the great-ape societies. Yet when we consider modern people in virtually all industrialized societies, we see status and hierarchy as the rule of life. David Erdal and Andrew Whiten [(1994, 1996)] of the University of St. Andrews in Scotland have depicted the evolution of hierarchy as a U-shaped curve, with the early rise of hierarchical society followed by the drop in status seeking among early humans. Hierarchical behavior returns only recently among modern people, perhaps when subsistence foragers became settled as agriculturalists and began living in larger groups. In the egalitarian relationships that exist among foragers, we may see the shaping of the origins of the human mind. Alliance formation, always important among nonhuman primates, assumes even greater weight among humans because of the value in cooperating to obtain meat."

(DV) [The real value (not understood by Stanford) is in cooperation, such as that in obtaining (particularly bigger game) meat, but also in other things, such that in the highly potentially fatal weaponized fighting between homonin groups increasingly invulnerable to predation due to improving weaponization and therefore increasingly shorter of supplies of more naturally preferred foods, members of a group are so tightly bonded that they will fight to the death, even if in retreat or flight, with fully committed cooperation. Also, the urges for competition and dominance within the group have an outlet in the hunting – the prey is the subject for domination and defeat instead of one's group-mates, and success in the hunt also serves as a proxy display of martial virtue or combat power – witnessing each other's lethal effectiveness in killing prey as large and larger than themselves is a constant reminder, for a cognitively capable hominin, that combat with each other is much more potentially fatal than it would be in the unarmed state. However, and very much to the point, it needs to be borne in mind that the urges for competition and dominance within the group have another outlet, in defense from predation – the predator's attack is the subject for domination and defeat instead of one's group-mates, and success in the defense also serves as a proxy display of martial virtue or combat power – witnessing each other's lethal or potentially lethal effectiveness in defending against predators as large and larger than themselves is a constant reminder, for a cognitively capable hominin, that combat with each other is much more potentially fatal than it would be in the unarmed state – and this would have come in early in hominid evolution, as use of clubs very probably occurred very early on. And there is a need for cooperation in defense against large predators, where a cognitively capable hominid in a small group (such as would be the theoretically accepted foraging party size in a mosaic environment of sparsely distributed resources, as it is in the great apes) might recognize that if his group-mates died from lack of support in defense from predation that he himself as a member of the continually reducing group would eventually likewise succumb. Mitigating against this would be the opportunity to join other hominid groups, but maybe more so at the earlier period in time when predation not competition between hominids was the limiting factor – at a later period in time, with increased size and strength, better biomechanics for upright activity including the better use of better developed weapons, and better cognition, predation was no longer the limiting factor to survival, but competition for limited preferred and even any food, and the option of joining another hominid group was no longer so easily available, particularly if one's cheating, battle-shirking behavior had been noted and remembered by other cognitively able hominid group members, who might not see such a potential group member as a positive acquisition for the group, the opportunity to join other groups after having become known by them to have let one's original group members down in encounters of either or both defense against predators and combat with rival hominid groups might plausibly have been much less available. These situations plausibly give rise to a selection for survival of greater and greater sharing behavior, and reduction of cheating behavior.]

(CS, THA) "Such coalitions in chimpanzees allow males jointly to challenge those much higher in the dominance hierarchy. Christopher Boehm of the University of Southern California suggests that in the course of human evolution, the political savvy that stems from the need to form alliances contributed to the rise of human intelligence. In humans, the ability to destabilize dominance hierarchies was the basis for the system of dominance reversal that characterizes hunter-gatherers [(DV) /gatherer-hunters]."
(DV) [And, at least as crucially the point, which Stanford unfortunately does not consider, potentially dominant group members might deliberately curtail their dominance behaviors themselves in order to ensure that the group will fight potentially lethal battles with other hominin groups successfully because the group will be bonded together in such a way as to make it likely that they will fight alongside each other in a fully committed cooperative fashion.]

(CS, THA) "Egalitarianism is a system in which the self-interests of the individual are at times submerged in order to benefit the group. …… So the goal is not to get ahead oneself, but to ensure that no one else gets ahead."
(DV) [And, at least as crucially the point, which Stanford unfortunately does not consider, to ensure that no one else falls too far behind, otherwise the group will not fight potentially lethal/ fatal battles with other hominin groups successfully because those who have been allowed to fall behind will not be bonded to the others in the group in such a way as to make it likely that they will fight potentially lethal/ fatal battles alongside them in a fully committed cooperative fashion.]

(CS, THA) "If the social intelligence required to navigate a social hierarchy is considerable, the political cleverness required to live within a nonhierchical system may be even greater. One must be acutely aware at all times of one's own needs, gains and losses, and be able to compare them to those of many other comrades."

Detailed review of The Hunting Apes
Chapter 7 Meat's Patriarchy (p 199-217)

(CS, THA) "(p202-3) In many traditional human societies, men hunt but women procure most of the protein and calories for their social groups through their gathering of roots, fruits, and small animals."
(DV) [Stanford fails to mention legumes, nuts, grains and other (i.e. oily) seeds, which is not a trivial failure, here...]

(CS, THA) "This behavior has led some anthropologists to claim that the importance of hunting and meat eating was more mythical than real, since men's efforts, while it received much attention from a previous generation of researchers, did not account for much nutritionally. But this misses the point of meat eating. The fact that meat is so highly valued even when it composes a small part of the diet is powerful testimony to its value as social currency. Men are able to use meat to enhance status, show beneficence, and even to obtain more sex by having caught meat. Nutrition is not irrelevant to meat's value at all; it is essential in that it makes the meat something worth bragging about, begging for, and manipulating with. ...... The value of meat is a matter of perception by group members. It is irrelevant that plant foods are as valuable a resource as meat. What matters is that animal carcasses are considered by both men and women as a more valued resource."
(DV) [See my comment above (p 91-92) for p196-8]

(CS, THA) "(p205) Competition among males, and female choice of male mates with particular physical features, led to the 10-15 percent size difference between ancestral males and females. Two biological anthropologists who are both feminists and sociobiologists, Sarah Hardy [(1997)] and Barbara Smuts [(1995)], have been exceptional in offering biological levels of explanation for human gender relations. In separate studies, both suggest that the same Darwinian principles that govern the conflicting interests of the sexes in all primate social systems may prescribe male coercive behavior toward females in humans. (CS, THA) ...... (p206) Females are not passive receptacles of male reproductive desires. They are active strategists in pursuit of their own interests and often are the driving forces between the social system itself." (DV) [The last part of this seems an overstatement of the reality – male domination, even if modified somewhat by females, seems more accurate a description.]

(CS, THA) "(p206) If patriarchies are about male dominance and the control of female reproduction, then we should consider how males acquire power. This may point to the role that meat eating and meat sharing plays in patriarchal systems. One way this occurs in nonhuman primate groups, human social groups, and probably in our ancestors' social groups, is through sheer size and strength. Sexual dimorphism is the product of natural selection favoring male attributes;"
(DV) [Incorrect, it is the product of natural selection favoring larger size and strength in males.]

(CS, THA) "...... (p207) A second way that males obtain power is through strategic alliances. This is true for both human and nonhuman primate societies. Male coalitions may form to control access to females, or to prevent other males having access to them, or to defend territories that contain desired females. ...... (p208) When it is highly beneficial for males to form alliances, there is often a kin-selected benefit that heightens the potential rewards for cooperating with kin. ...... Because female apes usually disperse from their natal groups at sexual maturity, they end up living as breeding adults in a new group lacking relatives and close allies. Some degree of bonding among females may develop over time, but females that disperse to new groups usually do not have a support system akin to that of males remaining in their natal groups. Barbara Smuts has pointed out that this lack of alliance formation contributes to the formation and maintenance of patriarchies by preventing females from bonding for purposes of protecting themselves from male ambitions. Females may, however, choose not to form alliances with other females because to do so would be against their own reproductive interests. ...... (p209-10) Meat is a difficult-to-obtain resource in both human and many other primate societies, valued out of proportion to its nutritional worth. ...... In her book The Sexual Politics of Meat, feminist writer Carol Adams [(1990)] considers the role of meat eating in human societies in shaping perceptions of women. She observes that women in many societies eat a diet that is mainly plant foods while men in the same societies eat more meat. The delegation of a [(DV) supposedly] second-class diet to women is one aspect of a patriarchal society. Peggy Sanday [(1981)] surveyed a range of traditional societies and concluded that those that eat a diet in which meat plays an important part are more likely to be strongly patriarchal ...... (p210) Adams points out that when males control a resource as highly valued as meat, its worth as a nutrient is largely mythical. Tubers and beans make an equally protein-rich diet. But men eat meat in many cultures in the belief that it gives them the strength that they need for their work. Women are not deemed to require it. Meat is also a nearly universal symbol of masculinity, from Western industrial to forager societies, and the eating of meat is thought to enhance masculinity. The image of a vegetarian weightlifter or football linebacker is paradoxical to most of us – evidence that these values are deeply entrenched in Western culture. ”

(DV) [Notwithstanding Dave Scott having won the Hawaii Ironman Triathlon (3.8km swim, 180km bicycle ride, 42km run), the world championship of such triathlons, six times as a vegan.....]

(CS, THA) “ Perhaps this is because meat eating is associated with meat getting and the other masculine attributes traditionally connected to hunting.. ...... Meat has long been a symbol of masculinity only because it served males well throughout human evolution as a political currency that is used to enhance male alliances, snub rivals, and control females.”

(CS, THA) “(p213) Female primates can navigate the complex social web of relationships, male power struggles, and their own power struggles only by being politically clever. ...... Being fertile, high-ranking, and clever are three important ways that females can obtain meat. Since no female is fertile for more than a small portion of her life, and only a few males can be high ranking, most must rely on their ability to manipulate their social milieu, including the males themselves, at the same time that males are trying hard to manipulate them."

(CS, THA) "(p214) The relationship between hunting and sex is nowhere clearer than among the Sharanahua, a people of the Peruvian Amazon Basin. According to anthropologist Janet Siskind, "Put at its crudest, the special hunt symbolizes an economic structure in which meat is exchanged for sex." Since Women produce as much food as men do, this is not likely to be just an economic exchange in which women exchange what they produce for what men produce. It is a culturally mediated system, with an obvious biological effect, in which sex is the incentive for men to hunt, and men who are better hunters have a better chance to have wives or mistresses. Siskind considers women to be a scarce commodity to be competed over in the Sharanahua, because sex is not free for men and must be won, and because some men have more than one wife, enhancing the impetus to "win" women with proffers of meat.

(DV) [But doesn't this, that some men may be not merely without a mistress but without a wife, argue against the theory of pure or nearly pure egalitarianism in hunter-gatherers (/gatherer-hunters)? Is what is happening is that meat is distributed in a fully egalitarian manner in order to set up not merely covert or tacit inequality in sexual access distribution but explicit inequality in sexual access distribution?

Elsewhere Siskind (1973) is cited as noting that "The *Sharanahua* have three ways of making women scarce: through limited sexual access, polygyny, and female infanticide.". But if women generally gather more than their own nutritional requirements, and men want more sexual partners, why kill them off? Limited food? In the Amazon there is generally accepted to be fierce and fatal wars between neighboring groups (i.e. the Yanomamo) - is there overall more women than men, in the absence of female infanticide, as men are killed off in warfare while women are not? Is the past different in this respect to the time Siskind studied them? Was there in the past enough infanticide biased against females to outweigh the war death biased against males? Or has the warring decreased, such that less men are killed off, and members of a tribal group do not need to be so strongly bonded as before to survive in a time of a lower rate of lethal warfare? Does Siskind mean “scarce” only in the sense of men always preferring to have more than one woman to have sex with, rather than in the sense of men outnumbering women?

In considering the representativity of the Sharanahua at the time Siskind studied them, we must be cautious about assuming that their social structure was representative of them at earlier times. In this regard note the following extract from Hern’s (1991) work on Amazonial tribes in general (The bolding of text other than headings has been added by myself):

“Generalizing from these sparse data is hazardous, but it seems that following a **catastrophic depopulation in the first 400 years following Western contact**, at least some Amazonian populations are experiencing high fertility and rapid population growth, whereas others have become extinct or nearly so. It remains to be seen whether those experiencing rapid population growth can maintain their traditional cultures in any respect.

In many ways, native Amazonians have experienced a reversal of the "epidemiologic transition" described by Omran (1971). Whereas Omran's theory begins with an "Age of Pestilence and Famine" and proceeds to an "Age of Receding Pandemics" and thence to an "Age of Degenerative and Man-made Disease", the Amazon Indians have had a mirror image of this experience since European contact. Roosevelt (1989) and Greene (1986) describe few except chronic diseases in *Marajó* skeletons, which preceded European contact. The pandemics began when the Europeans arrived. Coimbra (1989), in describing the health effects of disruptions of social networks and substistence patterns among the *Suruí*, shows that the group is experiencing all three phases of the "epidemiologic transition" at once.

HEALTH STATUS OF NATIVE AMAZONIANS

Our information about the health status of early Amazonians is almost nonexistent except for some skeletal remains found at *Marajó* (Roosevelt, 1989). There is evidence that tuberculosis, hookworm, *Trypanosoma cruzi*, and treponemal diseases were present in prehistoric South American populations (Allison, 1973; Coimbra, 1988; Baker & Armelagos, 1988; Hackett, 1963; Cockburn, 1961; Allison et al, 1974). However, most of the evidence for these diseases comes from West coast archeological remains, and we do not know the extent to which these diseases affected those in the lowlands. These afflictions are not reported in the *Marajó* individuals. In the rest of the Amazon, paleo-pathological evidence is lacking due to the unsuitable conditions for the preservation of human remains (Coimbra, 1988).

We must surmise a fundamentally vigorous and ingenious aboriginal population, however, which survived in a exceedingly complex and hostile environment by identifying a wide variety of food, technical, and medicinal resources (Berlin & Berlin, 1978; Dufour, 1987; Posey, 1983; Tournon, 1984; Hern, 1976; Milton, 1984; Ross, 1978; Behrens, 1981; Pollock, 1988). **The simple fact is that Amazon societies, however vigorous and complex, were overwhelmed by the introduction of Old World diseases and Europeans aggression**. Black (1990) asserts that the severe impact of epidemic disease introduced from the Old World was due both to social disruption and to genetic isolation.

INTRODUCED DISEASES

Measles (rubeola)

Black et al (1982) states that the measles virus probably did not exist at the time the Western Hemisphere was populated across the Bering land bridge, and probably dit not even exist at the time of European contact. His basis for this assertion is that Amazonian tribes having first contact typically display no evidence of exposure to the measles antigen, nor do they have antibodies to many other common viral disease agents. The antibody response of Ameridians to measles vaccination is essentially the same, if somewhat more symptomatic, as other populations, indicating no genetic immunodeficiency in this regard (Neel, Centerwall et al, 1970; Black et al, 1982; Black et al, 1970; Black et al, 1974).

Smallpox (Variola major)

Smallpox has been a great killer in Amazonia since the time of Conquest. From the time of the first reported epidemic of smallpox in 1621 (Meggers, 1971) until the last in 1964 (personal observartion), smallpox probably killed millions of native Amazonians. ...... Thornton et al (1991) have shown that native American populations may have had variable recovery rates following decimation by epidemics such as smallpox.

Tuberculosis

Tuberculosis has been endemic in the Americas since prehistoric times, and there is a high incidence of the diseases among contemporary Amerindian populations (Clark et al 1987). The introduction of tuberculosis into the Amazon is fairly recent (Black, 1975), but it is the main health problem in many native Amazonian groups today (Flowers, 1983; Black et al, 1974). Clark speculates that the current epidemic may stem from either exposure to more virulent strains of *Mycobacterium tuberculosis* or environmental change leading to a loss of exposure to natural vaccination, presumably when Amazonians were separated from their hemispheric ancestors. The widespread Amazonian practice of communal eating and drinking has unquestionably contributed to the epidemic spread of tuberculosis. In the Peruvian Amazon, in particular, the preparation of *masato* from masticated manioc is a popular and probably deadly custom in this regard.

Yellow fever

Cockburn (1961b) stated that the principal vector of yellow fever, *Aedes aegypti*, has probably been in South America for only 400 years. Black found high frequencies of antibodies to arboviruses, including yellow fever, among three Carib and four *Kayapó* Indian villages (Black et al, 1974), indicating prolonged exposure to these agents.

Chaga's Disease and Leishmaniasis

Cutaneous and mucocutaneous leishmaniasis is widespread throughout the upper Peruvian Amazon basin (personal observation), but documentation of the distribution of these diseases in Amazonia is sparse. Coimbra (1988) notes that *Trypanosoma cruzi* are found in Chilean mummies dating from 470 b.C. and that both the organism and its vector, the triatomina insect, are found throughout the Amazon basin. ......

In Coimbra's view, mobility is of adaptative value among preindustrial Amazon populations and minimizes contamination of surroundings with pathogens. Also, he notes that animal domestication was not generally practiced until recently.

Onchocerciasis

Microfilaria are found in various Amazon Indian groups (Lawrence et al, 1979; Beaver et al, 1976; Lawrence et al, 1980) and exposure to onchocerciasis has been found in Ecuador (Guderian et al, 1987) and Brazil (Moraes et al, 1974). While clinical manifestations of the disease have not yet become widespread (Salzano & Neel, 1976), it is only a matter of time before this severely debilitating disease affects large numbers of native Amazonians.

Schistosomiasis

Schistosomiasis has become widespread throughout the hemisphere but has not yet become in the Amazon. It is endemic in Northeastern Brazil (Lee, 1985:69), and it threatens to become a major health problem for Amazonians. The Amazon habitat is highly suitable for the spread of Schistosomiasis, especially if large dams are constructed in the region.

Helminthiasis

Infection with multiple species of intestinal parasites is the rule in relatively undisturbed Amerindian tribes, according to Lawrence (Lawrence et al, 1980). When native Amazonians become sedentary, however, they may be at higher risk of clinically important parasite burdens. Chernela and Thatcher (1989) found that the nomadic *Máku* had ascaris infection rates as low as 34%, whereas the 75 to 100% of sedentary *Tukáno* were infected. Schwaner and Dixon (1974) found that hookworm egg counts were six times higher in a sedentary unacculturated *Tukúna* Indians and among a mixed *Tukúna-mestizo* population living in a more urbanized settlement.

Chronic Diseases

The incidence of cardiovascular disease has not been found to be high in native Amazonian populations. ......

Donnely et al (1977) found that dental deterioration was positively associated with exposure to Western culture.

Violent and Accidental Death

In their study of the *Waorani*, Larrick et al (1982) found that 4% of all deaths were due to snakebite. While I have not calculated cause-specific death rates among the *Shipibo*, I have observed an

increasing proportion of deaths due to accidental gunshot wounds, homicide, and vehicular accidents since 1964. Reports of children drowning and dying from falls are common among the *Shipibo*.
Intervillage and intertribal warface has been an important cause of mortality for Amazonian tribes in the past, and Chagnon (1988) reports that 30% of all adult male deaths among the *Yanomámá* are due to violence.
Infanticide is still practiced in some Amazonian tribes, but its significance as a major cause of death appears to have diminished, especially in the past century."   ]

(CS, THA) "(p214-5) Machiavellian intelligence applies to both males and females, but it does not necessarily apply to them in the same ways. Males and females have been likened to separate species, with separate reproductive strategies and different means of meeting the same end of reproducing themselves. They may also have evolved somewhat different sorts of cognition as a result of needing to cope differently with complex social environments."

(CS, THA) "(p215) There is no reason to think that for several individuals sitting around a fire, meat holds the same promise. For a woman it may mean essential amino acids to nourish her fetus or infant,   "
(DV) [in times of a shortage of other foods, it may mean that, but otherwise, and also in any case, it is just as or more likely to constitute a mating display of martial and skill prowess.],
"   but to a man it may signify an ability to demonstrate his beneficence   "
(DV) [less so, as his sex has allocated the food gathering that is more of a drudgery to her sex]
(CS, THA) "   or his hunting [(DV) as proxy for martial] ability [(DV) more so] by sharing it with her. He may seek an alliance with her that is either politically strategic or reproductively beneficial, or both. She, meanwhile, has her own set of self-interests about control, reproduction, and political power [(DV) and of protection for herself and her offspring]."

(CS, THA) "(p216-7) I have portrayed the roots of human behavior as manipulation and social cunning that arise from the use of meat in our ancestors. This is very different from saying that, because of our meat-eating past, we have an innately aggressive nature. It should be remembered that predators, while possessing many adaptations that can be used in aggression within their own species, do not necessarily use them. In fact, most scholars doubt whether the predatory aggression that allows a lion to bring down a zebra or a chimpanzee to savagely rip apart a monkey is even closely related to intraspecific aggression. Aggressive behavior within a species may resemble predation, but the resemblance is often superficial."

(DV) [It needs to be remembered that often the only or major constraint on intraspecific violence is that being of the same species the combatants would be more evenly matched than predator and prey are, and therefore serious fighting is more likely to incur, even to the victor, a degree of injury that could threaten longer-term survival. And this is more so for animals that are obligate carnivores and are therefore equipped with more potently damaging weapons like claws and fangs and body musculature to go with the use of those in a combat situation – in animals less so equipped intraspecific violence is often greater in effort, as superiority in strength can be established with a lower risk of life-threatening injury.]

(CS, THA) "Humans are not demons by nature; in spite of the attention we focus on human violence, there are thousands of acts of compassion for every act of physical aggression."
(DV) [One obvious factor that Stanford has not addressed here is the difference between actions directed at members of another group, and actions directed at members of one's own group – evolutionarily speaking, and particularly in the case of small-group hunter-gatherers (/gatherer-hunters) it should be fairly obvious that these are likely to be two very different things.
Even a basic knowledge of human history should provide ample evidence for the substantial capacity of humans to inflict violence on members of other tribal, ethnic, socioeconomic, national, religious and gender groups.

Furthermore, that the capacity for sharing with members outside our own group, or outside groups that have (more tangible or obvious) mutually-beneficial relationships already, is very obviously very limited, otherwise there would not be the great disparity in socio-economic status that there is among the nations of the world, and within the nations of the world, such that what sharing there is could often reasonably be considered very token. Also there is the matter of psychological aggression, which although not directly causing death in the same way as physical violence does often seriously damages the recipients experience of life.]

(CS, THA) "The hunting, scavenging, and sharing of meat were fundamental features of the lives of our ancestors."
(DV) [But this varies across our evolutionary history, with, for example, the greater incidence of these during the end of the last glacial period of the last ice age (30,000 to 10,000 years ago) not being much representative of the much longer period of 6-odd million years before that.].
(CS, THA) "This does not mean that we are biologically driven to do any of these. The way that we deal with one another in society is rooted in social strategies that were molded during a time in our history when getting and using meat was vital."
(DV) [But mostly socially rather than nutritionally.]

(CS, THA) "If meat were a currency with a 10,000-generation history in the human family, then the traditions that have developed related to the use of meat are likely to have some evolved basis. By sharing meat we are both altruistic and selfish, as we are in most other arenas of our endeavors."
(DV) [Some consider that altruism is ultimately a nonsense concept, as we tend towards doing those things that we have been genetically or psychosocially programmed to, because we would feel uncomfortable to do otherwise, and this is worth keeping in mind, even though some 'meaning' in life might be experienced from the reduction of the suffering of those beings (of whatever species) that are suffering the most, even when other experiences such as material wealth and physical pleasure may be deemed

'meaningless'.]

(CS, THA) "We are not simply compassionate by training and Machiavellian by nature. Nor are we constrained by our past to repeat Machiavellian patterns in the future. A fuller understanding of our ancestral nature is, however, the first step to liberation."

References for review of The Hunting Apes:

Bailey DH , Geary DC (2009) Hominid Brain Evolution. Testing Climatic, Ecological, and Social Competition Models. Hum Nat 20: 67–79.

Hern WM (1991) Health and demography of native Amazonians: historical perspective and current status. Cad. Saúde Pública vol.7 no.4 Rio de Janeiro Oct./Dec. 1991

Kortlandt A (pre-1984 citations made in) Chapter 5 Habitat richness, foraging range and diet in chimpanzees and some other primates. In, Chivers DJ, Wood BA, Bilsborough A (1984) Food acquisition and processing in primates. Plenum Press, New York.

Leonard WR, Robertson ML, Snodgrass JJ (2007) Chapter 18 Energetic models of human nutritional evolution. In Ungar PS (Ed) Evolution of the human diet. The known, the unknown, and the unknowable. Oxford Press.

Lieberman P (2007) The Evolution of Human Speech. Its Anatomical and Neural Bases.
Current Anthropology 48 (1) February: 39-66 (Abstract only read)

Oftedal OT (1981) Milk protein and energy intakes of suckling mammalian young. A comparative study. PhD Diss. Cornell U, Ithaca, New York.

Vance DJ (2010/11) The Natural Human Diet? The Optimal Human Diet? The known and the unknown, the possible and the probable - an analysis of the evidence. Multifactor Health and Education Initiative (Multifactor Health website: http://sites.google.com/site/multifactorhealth)

Yamakoshi G (1998) Dietary response to fruit scarcity of wild chimpanzees at Bossou, Guinea: Possible implications for ecological importance of tool use. Am J Phys Anthropol. 106, 283-295. Cited by Lambert JE (2007) Chapter 17 Seasonality, fallback strategies and natural selection. In Ungar PS (Ed) Evolution of the human diet. The known, the unknown, and the unknowable. Oxford Press.

Right of reply offer to Professor Stanford, and any reply provided by him:

The author of the book critically reviewed here, Professor Craig B Stanford, was provided with a copy of this work above in mid-August 2010, and the only response he provided, was as just below.

As in the offer made at that time, if Prof. Stanford responds further at any time hereafter, including any indication of intent to respond at some stated or unstated time in the future, it will be added to this book at the Multifactor Health website, verbatim as provided, within 2 weeks of my receiving it.

Any response from me, here, to such response from him will likewise be communicated to him within 2 weeks of my making it.

And any response from him to that response, likewise.

And so on until all novel points, rebuttals, arguments or theories have been included accordingly here.

Prof Stanford's response, in full, was:

"I'm sorry, but given that I wrote that book more than 12 years ago (why would anyone be reviewing it now ?) and that there are some nonsensical comments in the review (weaponization in earliest hominids?.....), I don't think a reply from me is warranted."

My response to that was:

"Ok, that's what I'll put then.

I'm writing something larger; other people will still read your book; the matters are still debatable, and debated - ergo, I review. Furthermore, I've not seen potentially important matters I've elaborated on dealt with properly elsewhere.

My only response to your response will be that if chimpanzees can use hammer and anvil stones, and mortar and pestle stones, then early hominids can use clubs - hardly much of a stretch of the imagination - the combat reality is probably that they wouldn't have a hope of surviving without them anywhere other than very near climbable trees.

If you have anything else you want to add in response, let me know, and I'll add it verbatim."

And nothing further has yet been received from Professor Stanford.

Detailed Critical Review of:

Catching Fire How Cooking Made us Human.

Richard Wrangham, professor, Harvard University
(2009 Basic Books, Perseus Books Group.)

Contents:

Abstract for critical review of Catching Fire:

Professor Wrangham's theory is essentially that the modern human gut (digestive tract), dentition (teeth), jaw bone and musculature size and strength, brain size, body size, and furthermore much of modern human basic behavioural instincts and practice, are the direct result mostly of the eating of cooked instead of raw food.
In particular, compared to our nearest animal relatives, the apes, and what is surmised about our pre-human ancestors and the common ancestor(s) of humans and the apes:
the modern human stomach and large intestine being smaller; and the small intestine being longer;
the modern human teeth being smaller;
the modern human jaw bone and musculature being smaller and weaker;
the modern human brain being bigger;
the modern human size being larger;
the modern human tendency to share food socially being greater; and the modern human tendency for longer-lasting and more exclusive female-male pairs being greater.

This(ese) evolution(s) are claimed to be due to cooking:
increasing the digestibility of foods;
increasing the range of food being edible;
involving campfires that allow sleeping on the ground instead of in the trees;
requiring the sharing of food;
requiring and facilitating the division of labour between the sexes;
reducing the spacing of offspring by allowing earlier weaning; and increasing the value of input (such as of cooked food) from grandmothers

While he is correct inasmuch as cooking almost certainly does increase the digestibility and therefore the energy available from a large proportion of foods including those eaten during the course of human evolution from our common ancestor(s) with the ape(s) (chimpanzees, most recently), (though he fails to note the attendant increase in dental decay, and to address the aspect of lowered stability of blood sugar and lipid levels), Wrangham's theory is in my opinion fatally deficient, such as to be near enough to fully incorrect in all other major points, due to:

his failure to consider that the increased consumption of seeds (cereal, and oily) and nuts, generally among the most energy-dense of foods, continued or increased consumption of legumes, which are highly energy dense, along with increased consumption of starchy root vegetables, along with the processing of these and other foods by cracking, grinding, crushing, mashing, peeling, drying and soaking using naturally formed stone and wood (and other) tools, in conjunction with storage and transport, is at least as likely as the eating of cooked food to have resulted in those changes to the human anatomy, as well as to human behavioral instincts and practices;

his taking to be fact that modern humans cannot derive enough energy from even the most optimal choice and processing of raw foods, which is far from true;

his taking to be the time of the inception of cooking as a general practice is very presumptively towards a time far earlier in human evolution than the evidence would suggest, the evidence regarding this not having been even close to properly analysed;

his selection of times of sudden change in the course of evolution of (pre-)human brain size being grossly presumptive as there are simply not even close to enough fossil remains to evidence the sudden changes that are a major basis of his theory;

his failure to have properly considered that archeological discovery biases exist, in particular that drier areas that were population sink areas (not population generating areas) for ancestral hominins are much more likely to provide archeological remains than moister areas that were population generating areas; and that tools and the bones of large animals as archeological remains are more likely to trigger the discovery of any associated hominin remains than the hominin remains themselves; and that

animal bones remain while plant food refuse does not as it is much more susceptible to decay;

his failure to have properly addressed the fact that very early in evolution our early hominid ancestors lost the elongated canine teeth that chimpanzees still possess and use not only for the combat that is most of their use but also for what little meat eating they do, never regaining them throughout the subsequent 6,000,000-odd years of evolution up to and including the modern human, whereas almost all animals with any significant meat eating in their diet have elongated canine teeth or some functional equivalent of them; this in conjunction with his failure to have addressed at all the fact that modern humans have thicker and stronger dental enamel than any of our nearest animal relatives the apes;

his failure to properly consider the likely evolved taste preferences of chimpanzees, considering that they do not actively set out to hunt meat, taking it only when they come on it, and even then usually not in seasons of fruit plenty, rather when they are in some level of starvation (the other apes even less so, that is, almost nil); and his failure to relate this to the likely evolved taste preferences of hominins; and his failure to relate these to the fact that the use of fire and cooking very often changes naturally unpalatable foods into palatable foods (in conjunction with the facts of habit acquisition), and the possibility that raw meat is, and was in our evolution of the last 6,000,000-odd years, one of those naturally unpalatable foods; in conjunction with,

his failure to have examined at all the appropriate biochemicals likely to provide evidence of the proportion of diet that meat constituted during the evolution leading to modern humans:
vitamin C biosynthesis, unlike nearly all other animals, lost through evolution;
vitamin B12 absorption, transport, and storage, in sharp contradistinction to vitamin C, retained in high function through evolution;
human dietary protein requirements being so low, as evidenced both by adult requirements and human breast milk protein content

relative to energy content being only a fraction of that of all other mammals known;
human water requirements and the ability of fruit and vegetable food to meet the large majority of them, as in apes and monkeys;
50% of dietary calories as protein being overtly toxic, diuretic, attracting predators by odor, and increasing the cost of digestion;
meat being less energy-dense than nuts, seeds, grains and legumes;
human starch digestion having been enhanced in evolution by the triplication of salivary amylase;
taurine being very likely sufficient in even vegan human breast milk;
elongation and desaturation of 18 carbon fatty acids to their product 20 and 22 carbon fatty acids being very likely sufficient as evidenced by even vegan human breast milk fatty acid contents;
vitamin A synthesis from beta-carotene being retained through evolution, in contradistinction to carnivorous animals;
adult lactase persistence in humans being present in only a small fraction of the world's human population;
fructose metabolism being very well conserved in all human populations through evolution;
gluten allergy/ celiac disease not being more prevalent proportional to consumption than allergies to significant animal foods;
all of which, and with virtual absolute certainty when considered together, indicate that the proportion of diet that meat constituted during the evolution leading to modern humans must have been miniscule (other considerations strongly suggesting that what proportionately little animal matter was eaten was mostly insects, with occasional eggs) (see also p 178-183 below);

his failure to properly consider the likelihood of increasing weaponization of hominins from even the earliest stages of evolution from our common ancestor with the chimpanzees;
and his failure to relate this (at all) to both the likely intra- and inter-group defensive needs of hominins of all stages of evolution as they become increasingly lethally weaponized;
and his failure to relate these (at all) to the likely changing proportions of need of hominins, as they become increasingly lethally weaponized, for defense from predators versus from other

hominins;
and his failure to relate these (at all) to the probability that hunting can be a proxy display (both for mating and male-male competition/ dominance) of martial virtue that reduces serious damage to combatants of the same hominin group that would likely occur through the combat alternative, together with the probability that hunting can be an outlet for aggression naturally evolved to establish dominance in a primate group and reap the mating opportunities commensurate with that, reducing serious damage to participants in, or victims of, combat or attack, respectively, between/ among members of the same hominin group, together with the probability that hunting can be a proxy display of martial virtue that reduces serious damage to inter-group combatants that would likely occur through the combat alternative in the establishment and maintenance of territories of hominin groups, together with the probability that hunting constitutes an escape from domestic drudgery for males;

his failure to properly consider the role of defense requirements in the evolution of female-male division of labor;

his failure to properly consider the role of meeting the needs/ desires of males for regular, safe (from conflict), low-time- and low–effort-cost sex, together with the need for males of a group to be strongly bonded into a fighting group that will fight together in a strongly cohesive and committed manner, as being major determinants of the evolution of significantly exclusive female-male pairing;

his failure to properly consider the use of fire versus the use of sentries, in the provision of the safety of hominins sleeping on the ground, including the differential effect of fire in being a repellant of non-human predators while serving as a locater and visibility-enhancer for attacking rival hominins;

his underrating the climbing ability of both modern humans and our pre-human (hominid) ancestors, in conjunction with not understanding that nature almost always doesn't require the

survival of the aged and the infirm for the promulgation of the species;

his failure to have considered the role of the evolution of the modern human supra-laryngeal vocal tract (required for the formation of speech (as distinct from uttered sounds), which apes are not capable of) in the evolution of the smaller dentition and jaw bone and musculature of modern humans; compounded by a failure to consider the fact that the enamel of the teeth of modern humans is thicker and stronger than that of any of the apes;

his failure to have considered the significance of the fact that gatherer-hunters/ hunter-gatherers of modern and recent historical times live and eat in areas of the world that are more marginal (less rich in human foods) than areas that have been taken over by agriculture, in particular but not limited to the fact that grains would represent less of the diet of such groups because areas suited to grains were taken over by agriculture (largely the growing of cereal grains);

his failure to have properly considered the possibility that the evolution of the larger modern human brain was driven by multiple factors/ aspects, including as major factors: the survival benefit to hominins of being able to use an increasing complexity of food sources and food processing; the survival benefit to hominins of being able to incorporate an increasing strategic complexity in their intra-group and inter-group social interactions.

In conclusion, taking the above facts, factors and considerations into account, the only tenable conclusion is that Wrangham's main theory(ies) in his book "Catching Fire How Cooking Made us Human" are so badly flawed that they must be virtually completely discarded.

Detailed Review of Catching Fire

Detailed review of Catching Fire
Introduction: The Cooking Hypothesis (p 1-14)

(DV) [Wrangham lays in some prehistorical background including a fanciful scenario in which an australopithecine (one line of which we humans are descended from) in a modern sporting stadium:]
(RW, CF) "(p3) … grabs an overhead beam and swings away over the crowd to steal some peanuts from a vendor."
(DV) [Modern chimpanzees climb trees, but do not brachiate (moving through trees by swinging from the arms) as the monkeys do – and they usually walk on the ground between the trees that they climb (Teleki 1981, who also cites Napier and Napier 1967) . Therefore it is highly unlikely that australopithecines, with their feet much more evolved for walking than for climbing (quite proficient climbers that they nevertheless still were, for the purposes of harvesting some of their food and probably also for defense and occasional sleeping), would brachiate.]

(RW, CF) "(p3) Cut marks on fossil bones show that the simple knives [made from cobblestones being deliberately clashed, in Ethiopia, 2.6 million years ago] were used to cut tongues out of dead antelope and to get hunks of meat by slicing through tendons of animal limbs."
(DV) [Wrangham fails to note here that stone knives were also used for many other purposes than butchering animals, including processing plant foods and making various objects (Ungar 2007 (1) (cites also Keeley and Toth 1981), Walker 2007 (cites also Keeley and Toth 1981), Shea 2007, Donald 1991, Richards 2002 (though he too jumps to the wrong conclusion)). Furthermore, he fails to attempt what quantification of those purposes there is available from studies of later humans and tool use, and of evidence suggesting (albeit biased by the then climate(s) and ecology including regarding population density) what the proportion of diet that meat eating is likely to have been (as do, utterly, Blumenschine and Pobiner 2007, Bunn 2007, Cordain 2007, Eaton, Shostak and Konner 1988), which is very little, see my example calculation

from the Solutre data (from Olsen 1989) on p 236-237 below in my reviewing of Wrangham's Chapter 8.]

(RW, CF) "(p5)... modern humans emerged about two hundred thousand years ago."
(DV) [the fully modern human supra-laryngeal vocal tract, that allowed the full range and fully audible distinction of human speech, was not evolved at 400,000 y BP (Capasso, Michetti and D'Anastasio 2008), long after the advent of *Homo erectus*, and possibly not till as late as around 40,000 years ago – which is a not trivial point, given the relationship between intelligence and language (Donald 1991).]

(RW, CF) "(p5) So the question of our origins concerns the forces that sprung *Homo erectus* from their australopithecine past. Anthropologists have an answer. According to the most popular view since the 1950s there was a single supposed impetus: the eating of meat."
(DV) [Wrangham fails to note that even if at one time the Man the Hunter hypothesis was popular (Lee and de Vore 1968), that most experts in the field these days in 2010 and for many years before this time have regarded the Man the Hunter hypothesis, and this last sentence of Wrangham's above, as incorrect, or at the very least very much a conjecture, far from being conclusively established as or even strongly suggested to be, on the evidence, a fact (Panter-Brick 2002, Nestle 1999, Sept 2007, Potts 1992, Harding and Teleki 1981, Mann 1981 (cites also Jolly 1970, 1972, Isaac and Crader 1981), Hart and Sussman 2007 (though criticized by Muller 2007 for lack of thoroughness, though Muller himself fails to note the crucial lack of quantification and bias identification in archeology, and lack of examination of relevant nutritional biochemistry (see p 178-83 below) in contradictory/complementary work he refers to as overlooked or not addressed by Hart and Sussman while noting the ease of their "kicking around Ardrey and Dart" as early Man-the-Hunter theorists.))]

(RW, CF) "(p6) Hundreds of different hunter-gatherer [(DV) on average they are more correctly termed gatherer-hunters...]

cultures have been described, and all obtained a substantial portion of their diet from meat, often half their calories or more."
(DV) [Wrangham fails to note that virtually all of these historical gatherer-hunters/ hunter-gatherers live(d) on more marginal lands, because the richer lands were taken over by agriculture (Eaton, Konner and Shostak 1988, Walker 1981) – therefore their diets probably are more representative of fallback diets that prehistorical humans ate at times (seasons or periods) and in places when/where more naturally preferred foods were not available (Lambert 2007). And he fails to note that more gathering than hunting generally occurs, making 'gatherer-hunters' a more generally correct term than 'hunter-gatherers' (Hayden 1981).]

(RW, CF) "Archeology indicates a similar importance of meat all the way back to the butchering habilines more than two million years ago."
(DV) [This is quite incorrect – the archaeological evidence, and the large majority of the authoritative literature on it, indicates nothing of the sort (Potts 1992, Shea 2007; see calculations (p 236) on the Solutre data of Olsen 1989 in my review of Wrangham's Chapter 8 below; and note that Blumenschine and Pobiner 2007, Bunn 2007, Cordain 2007, Richards 2002, and Eaton, Shostak and Konner 1988 all plunge ahead into their assumptions of meat constituting a large proportion of our ancestral diet without so much as attempting the basic calculations from the evidence available of what that proportion might be; and note that Wrangham does not provide any citations here in support of this crucially important claim).
For example, Donald (1991, pp 111, 112-116) asserts: that habiline tooth microwear patterns do not indicate meat-eating (DV: though meat eating might not show well) ; that the enlargement of the human brain is associated with increasing sophistication of stone tool manufacture; that such stone tools were possibly more likely to be employed in the processing of plant foods and the making of objects not related to meat-eating; that the evolution of the modern supralaryngeal vocal tract involves marked changes in mouth cavity structure.
Furthermore, it is very well accepted that archeological remains are very selected, that is, not representative of the general past, because

they are biased, by animal bones being preserved whereas plant food refuse decays (Mann 1981, Delson et al (Eds) 2000), and biased by dryer climatic areas (where there was probably proportionately more carnivory than there was in moderate (moister) climatic areas) preserving more archeological remains than moister climatic areas, those dryer areas probably being population sink areas where our ancestors did more dying than reproducing and those moister areas probably being population generating areas where our ancestors did more reproducing than dying (thereby actually being the areas that our evolution mostly occurred in…) (Mann 1981, Peters 2007).
The habilines very probably only butchered occasionally, in times of great hunger, not least because they preferred the taste of other (i.e. plant) foods – "the butchering habilines" gives a very false impression of the probable extent of their carnivory…]

(RW, CF) "By contrast there is little to suggest that their predecessors, the australopithecines, were much different from chimpanzees in their predatory behaviour. Chimpanzees readily grab monkeys, piglets, or small antelopes when opportunities arise …"
(DV) [Usually they don't, and they do so more in seasons of scarcity of the fruit that is their more preferred food (Stanford 1998, despite his subsequent exaggeration and sensationalization of chimpanzee carnivory in his book of 1999, Kortlandt 1984 citing Kortlandt 1967, Teleki 1981 citing van Lawick-Goodall 1968, Yamagiwa et al 1996,), scarcity severe enough that growing juveniles stop growing, that is, some degree of starvation (Hladik 1981) …],
(RW, CF) "but weeks or months can go by with no meat in their diets."
(DV) [What evidence there is includes that chimpanzees use their elongated canine teeth to kill what prey they do eat – australopithecines did not have elongated canine teeth, and the fact that they lost the elongated canine teeth in evolution from the common ancestor of the modern chimpanzees and themselves fairly strongly suggests that in fact they ate substantially less meat than chimpanzees do (or did, which is not necessarily the same thing,

considering the possible effects of humankind's various impacts on the modern chimpanzees situation(s), including on their food supplies, predator (cats) populations, and imitatable role models.)]

(RW, CF) "(p7) So it is easy to imagine that the rise of meat eating fostered various human characteristics such as long-distance travel, big bodies, rising intelligence, and increased cooperation. For such reasons the meat-eating hypothesis, often called "Man-the-Hunter," has long been popular with anthropologists to explain the change from australopithecine to human."
(DV) [Again, Wrangham fails to note that even if at one time the Man the Hunter hypothesis was popular, that most experts in the field these days in 2010 and for many years before this time have regarded the Man the Hunter hypothesis as incorrect (references given above p7-8)…]
(RW, CF) "But the Man-the-Hunter hypothesis is incomplete because it does not explain how hunting was possible without the economic support gathered foods provided. … …
(p8) Meat eating accounts smoothly for the first transition [(DV) around 2.5 million years ago], jump-starting evolution towards humans by shifting chimpanzeelike australopithecines into knife-wielding, bigger-brained habilines, while still leaving them with apelike bodies capable of collecting and digesting vegetable foods as efficiently as did the australopithecines."
(DV) [This smooth accountancy is very much in the wishful imagination of Wrangham, rather than based on any evidence – and he cites no source of any such evidence in the 43 pages of notes at the back of the book, which is very much occupied with giving and further detailing his references for the book.]
(RW, CF) "But if meat eating explains the origin of the habilines [(DV) it doesn't (references given above)…], it leaves the second transition [(DV) around 1.9-1.8 million years ago] unexplained, from habilines to *Homo erectus*. Did habilines and *Homo erectus* obtain their meat in such different ways that they evolved different types of anatomy? Some people think the habilines might have been primarily scavengers while *Homo erectus* were more proficient hunters. The idea is plausible, though archaeological data do not directly test it."

(DV) [The last sentence here is plausible – what is not plausible is that meat was a large enough proportion of the diet to have much influenced our physiological evolution, as clearly indicated by the lack of evidence for it (references given above), and the quite conclusive biochemical evidence against it as I present in my reviewing of Wrangham's Chapter 4 below (Vance 2010, Ch 3).]

(RW, CF) "(p8) But it does not solve a key problem concerning the anatomy of *Homo erectus*, which had small jaws and small teeth that were poorly adapted for eating the tough raw meat of game animals."

(DV) [The jaws and teeth of the habilines were also poorly adapted for eating raw meat, looking nothing like those of the more obviously evolutionarily naturally carnivorous animals – what they are fairly clearly evolutionarily adapted at least in very large part for is eating a very predominantly vegetarian diet including some or all of fruit, legumes, nuts, oily seeds, cereal grains, other seeds, starchy root vegetables and green leaf vegetables.]

(RW, CF) "These weaker mouths cannot be explained by *Homo erectus*'s becoming better at hunting. Something else must have been going on."

(DV) [Indeed, including that they were not eating much meat… It is very defective that Wrangham has here failed to address the logical alternative(s). (Most of these references do not address particulars much, but clearly indicate the need for the consideration of the alternatives: Peters 2007, Reed and Rector 2007, Sept 2007, Lambert 2007, Lucas 2007, Ungar 2007 (1 & 2), Teaford 2007 (1 & 2), Daegling and Grine 2007, on various aspects of the consumption of other foods and/or tool use, and the quite conclusive biochemical evidence against there having been much meat eating, as I present in my reviewing of Wrangham's Chapter 4 below (Vance 2010/11, Ch 3))]

(RW, CF) "(p12) Edmund Leach presented Levi-Strauss's views crisply: "(People) do not have to cook their food, they do so for symbolic reasons to show that they are men and not beasts."" … …

(DV) [There is some important truth in Levi-Strauss's views, assuming that Leach has presented them here correctly, and it seems to me that these symbolic reasons, including also but not properly explored here technological and pseudo-magical skill status displays, are a major factor in the establishment of cooking as a nearly universal practice of humankind. However it is very deficient not to address the great probability that another major factor is cooking's improvement of the taste of foods that would otherwise be naturally, and at least to some extent probably optimally, not those having a taste preferred by humans, and therefore possibly more naturally fallback foods (Lambert 2007) eaten only in the absence (i.e. seasonal, or other-cyclical or –periodic) of preferred foods – for example improving the taste of raw meat, which is probably evolutionarily physiologically naturally less palatable than raw nuts, oily seeds, grains and other seeds as a protein source, or than these and raw fruits and starchy root vegetables as an energy source.

Note that it seems reasonable that less-palatable things that are easily obtainable are more likely to be by evolution a significant part of the diet than less-palatable things that are not easily obtainable – and for the very large majority of the 6 million years of humankind's evolutionary history since our line split from that of the chimpanzees, raw meat has probably been a less-palatable thing that was not easily obtainable, though rotting raw meat would have been more easily obtainable it was almost certainly less-palatable, the acquired and arguably perverted tastes of some very recent and present-day humans notwithstanding…]

(RW, CF) … … "His [Charles Darwin's] experience told him that cooking helps us to eat meat more easily. After our ancestors started cooking, he argued, meat became more desirable and valuable, leading to a new importance in hunting."

(RW, CF) "(p13) Cooked food does many familiar things. It makes our food safer, creates rich and delicious tastes, and reduces spoilage. Heating can allow us to open, cut or mash tough foods. But none of these advances is as important as a little-appreciated aspect: cooking increases the amount of energy our bodies obtain from food. …"

(DV) [Cooking does not always make food safer – subjecting food to the higher temperatures used in cooking, in the presence of oxygen, causes a large range of unhealthy chemicals to be formed, such as oxidatively damaged (C=C double bonds oxidized to epoxides and peroxides) unsaturated fats, Maillard Products (amino acids, notably lysine, bonded to a sugar, and then unabsorbable from the gut), polycyclic and heterocyclic products (mutagens/ carcinogens), and carbonized (charred, mutagenic/carcinogenic) material. Also it significantly reduces the levels of many important vitamins. Also it can both encourage and facilitate the overeating involved in the obesity that many populations now have very damaging epidemics of…]

(RW, CF) "(p14) The extra energy gave the first cooks biological advantages. They survived and reproduced better than before. Their genes spread. Their bodies responded by biologically adapting to cooked food, shaped by natural selection to take maximum advantage of the new diet."
(DV) [This is just Wrangham's conjecture. It is very incorrect to state this as a fact. There is not at all enough evidence to support this being acceptable as a fact – if one wants to introduce novel theories such as this (mostly disagreed with by others in the field) and have them viewed as even potentially somewhat credible one needs to present sufficient evidence to support them – Wrangham does not do so here in his book because such evidence is simply lacking in the annals of the known science – this is just another 'just so story', making up a theory story as one goes along…]

(RW, CF) "There were changes in anatomy, physiology, ecology, life history, psychology, and society. Fossil evidence indicates that this dependence arose not just some tens of thousands of years ago, or even a few hundred thousand, but right back at the beginning of our time on earth, at the start of human evolution, by the habilines that became *Homo erectus*."
(DV) [This is pure conjecture! There is not even close to enough evidence to support this claim being even potentially plausibly possibly factual, and that is why one will not find it within this book!]

Detailed review of Catching Fire
Chapter 1 Quest for Raw-Foodists (p 15-36)

(DV) [In Chapter 1 Wrangham makes (the whole chapter is) the quite false claim that humans cannot readily meet their caloric (energy) requirements from raw food, a claim of his that I have comprehensively demolished previously in my (2010) book "The Natural Diet of Humans? The Optimal Diet of Humans? The known and the unknown, the possible and the probable - an analysis of the evidence," available free at the Multifactor Health website (http://sites.google.com/site/multifactorhealth) (but surely better purchased in print…), and that I will deal with, in somewhat different and more complete detail, again here.]

(RW, CF) "(p16) … Evo Diet Experiment. In 2006 nine volunteers with dangerously high blood pressure spent twelve days eating like apes in an experiment filmed by the British Broadcasting Corporation. … ate almost everything raw. Their diet included peppers, melons, cucumbers, tomatoes, carrots, broccoli, grapes, dates, walnuts, bananas, peaches, and so on – more than fifty kinds of fruits, vegetables and nuts. In the second week they ate some cooked oily fish, and one man sneaked some chocolate. … … The participants ate until they were full, taking in up to 5 kilograms (10 pounds) by weight per day."
(DV) [In order to have meaningful evidence from this experiment, it is necessary to know what "up to 5 kilograms" means – we need to know the *average* intake, and also the *range* of intakes, not just the *maximum* intake.
In order to have meaningful evidence from this experiment, it is also necessary to know what the meal planning and eating schedule were – for example, if the more calorically-, fat- and protein-dense foods in particular the nuts were eaten at the end of a meal comprised mostly of the less calorically-, fat- and protein-dense foods in particular the melons, cucumbers and tomatoes, then there is a fair chance that a significant fraction of the nuts were not digested and absorbed as well as they would have been if these two dietary groups were eaten in the opposite order in the meal…
In order to have meaningful evidence from this experiment, it is

also necessary to know how well the food was chewed, which is closely related to how quickly or slowly the food was taken in – if the food was chewed in the usual fashion that the food usually eaten (cooked food, requiring little chewing) by most people who have significantly raised cholesterol and blood pressure is, it was probably far from properly chewed, and then, in a fairly substantially unchewed state, which is known to reduce the absorption into the body of the nutrients in the food eaten, the meal as a whole is eaten far too quickly for digestion and absorption of nutrients to be optimal, even if the food had been properly mechanically broken down (e.g. chewed, or by machine) and insalivated with oral digestive enzymes, because subsequent contact with other digestive enzymes and media is physiologically (including biochemical (e.g. hormonal) control) and chemically suboptimal.

The fact that Wrangham fails to so much as mention, let alone properly address, any of these crucial experimental issues clearly indicates that he is either ignorant of their crucial significance or glossing over them in the hope that most of his readership will be blithely unaware of them and so swallow down the matter so palatable to them (and Wrangham) due to their own cherished dietary habits.

It also suggests the experiment's nutritionist was similarly defective in their understanding of these crucial issues, therefore contributing to yet another (and they are many in number and quite common) scientific study that is nearly meaningless due to the flaws in its experimental design and data collection and analysis.]

(RW, CF) "The daily intake was calculated by the experiment's nutritionist to include an adequate 2,000 calories for women, and 2,300 calories for men." … …

(DV) [If this, and the little other mentioned by Wrangham immediately below, was the extent of the quantification and calculation, then as I note just above, this study was almost certainly yet another (and they are many in number and quite common) scientific study that is nearly meaningless due to the flaws in its experimental design and data collection and analysis. Other, that is, than as an illustration of what happens due to these design flaws, *if* they and their effects on the study results are properly elucidated.]

(RW, CF) "(p16-17) By the end of the experiment their cholesterol levels had fallen by almost a quarter and average blood pressure was down to normal. But while medical hopes were met, an extra result had not been anticipated. The volunteers lost a lot of weight – an average of 4.4 kg (9.7 pounds) each, or 0.37 kg (0.8 pounds) per day."

(DV) [This substantial fall in cholesterol over such a small time, along with an ostensibly significant fall in blood pressure, strongly suggests that the diets of the study subject-participants prior to the Evo Diet Experiment had contained quite a lot of added salt, which not only raises blood pressure, but also (which is mostly what raises the blood pressure…), causes the body to retain excess water – when the excess salt intake was stopped on the Evo Diet, that excess water would have been lost, which is very likely to have been more than half of the weight lost during the Evo Diet Experiment. More relevant here than the average loss of weight per day is the rate of loss in the earliest days and the rate of loss in the later days – which we are not provided with.

And again, The fact that Wrangham fails to so much as mention, let alone properly address, this crucial experimental issue clearly indicates that he is either ignorant of its crucial significance or glossing over it in the hope that most of his readership will be blithely unaware of it and so swallow down holus-bolus the matter so palatable to them (and Wrangham) due to their own cherished dietary habits.

It also suggests the experiment's nutritionist was similarly defective in their understanding of this crucial issues, therefore contributing to yet another (and they are many in number and quite common) scientific study that is nearly meaningless due to its experimental design and analysis flaws.]

(RW, CF) "(p17) Raw-foodists are dedicated to eating 100 percent of their diets raw, or as close to 100 percent as they can manage. There are only three studies of their body weight, and all find that people who eat raw tend to be thin. The most extensive is the Giessen Raw Food study, conducted by nutritionist Corinna Koebnick and her colleagues in Germany, which used questionnaires to study 513 raw-foodists who ate from 70 percent to 100 percent of their diet raw (Koebnick et al 1999). They chose to eat raw to be healthy, to prevent illness, to have a long life, or to live naturally. Raw food included not only uncooked vegetables and occasional meat, but also cold-pressed oil and honey, and some items that were lightly heated such as dried fruits, dried meat, and dried fish. … … the body weights of almost a third indicated chronic energy deficiency. The scientists' conclusion was unambiguous: "a strict raw food diet cannot guarantee an adequate energy supply."

(DV) [This conclusion is in fact somewhat all of ambiguous, meaningless, redundant, and plainly silly – if inappropriate (and I would call this plainly silly, from a nutritional scientist's point of view) proportions of nutritionally different food groups were eaten, for example nothing but raw cucumbers and tomatoes (which have approximately no energy value), of course an adequate energy supply could not be had; likewise if nothing but cooked cucumbers and tomatoes were eaten, an adequate energy supply could not be had.

Alternatively, if a person consumed (properly chewed, and in the correct order (as follows)) the 5 kg weight per day referred to in the Evo Diet Study description (see just above), in raw nuts and bananas (which are very high in energy value), unless they had something wrong with their ability to digest and absorb food, or they vomited it back up from the sheer excess, they would be gaining body weight at a very rapid rate indeed due to a *massive* excess of energy intake (at 2.5 kg of each, about 2,500 calories for the bananas plus about 15,000 calories for the nuts (see any standard nutrition content tables), total about 17,500 calories, or more than 6 times the estimated daily energy requirement). If anyone should doubt this, it is very easy to test it – just get the food and eat it and monitor the weight gain, if one could force oneself to

overeat like that… My experience with much smaller amounts of nuts or bananas in excess of my requirements, in readily adding (excess) fat to my carcass, is wholly in accord with my statements here.
Furthermore, in noting failure to meet caloric requirements it is necessary to differentiate cases of not only obvious and diagnosable/diagnosed anorexia nervosa, but also less severe cases on the continuum of (under)eating disorder that are not diagnosable as anorexia nervosa nor diagnosed as less severe cases of (under)eating disorder, from the whole group, or the analysis will be fatally flawed.
Having looked at the journal article abstract and a related journal article (the abstract for Koebnick et al 1999, and a 2005 journal article) and a website by the authors, it is obvious that there is no mention of the Giessen researchers screening out anorexia nervosa patients, and, there being in fact in the population a continuum from milder (under-)eating disorder cases through to more severe ones, that the Giessen methodology for acquiring their subjects would have not have screened out of the study group any other than the most severe staggering cases of (under-)eating disorder.
Furthermore, the association with underweight is further confounded by the fact that roughly 50% of the study subjects took up the raw food diet as a way of dealing with a pre-existing disease.]

(DV) [Furthermore, consistency of results among studies means nothing if all studies are making the same mistakes, which is not a very rare event, actually…]

(DV) [That various deluded and pseudoscientific uses of and claims for raw-foodism are indeed, as Wrangham notes (p23-25), made, is not an argument against the sensible eating of raw foods – every other animal on the planet does it, naturally, after all…]

(RW, CF) "(p25-26) Eating raw intrudes into social life, "
(DV) [True, though this can be a benefit as well as a liability, and educating others against their bad habits, misunderstandings, ignorance or delusions often carries a (transient, or long-term…)

penalty, and one has to decide one's own resolution of that tradeoff, which will differ at different stages of one's personal evolution and progress…]

(RW, CF) " demands a lot of time in the kitchen, "
(DV) [This need not be true at all – I myself spend very little time in food preparation, mostly just grabbing, cutting and eating – next to no time, and likewise for dishwashing, because a quick water rinse is all that is necessary. It is rather *cooking* that takes up time, perhaps obviously enough, if one wasn't making up fiction as one went along to try convince oneself of the tenability of one's cherished dietary habits…]

(RW, CF) " and requires a strong will to resist the thought of cooked food."
(DV) [At first, yes this is true – the habits of a lifetime are indeed hard to break – but no pain, no gain – as time goes by and the habits are broken and replaced with better ones, less and less effort is required – I myself, now a raw-fooder of more than 20 years, can truthfully state that the practice has been near enough to totally effortless for the large majority of that timespan. With a good/sufficient understanding of *why* what one is doing is a good thing to be doing, and a good/sufficient understanding of the involvement of one's own and others psychology in the matter, the old habits are soon enough broken and replaced with better habits…]

(RW, CF) "(p26) It can create personal problems, such as annoyingly frequent urination, "
(DV) [True, it can, mostly in winter (particularly in locations we are obviously not evolved to be best suited for…) when little sweating and less transpiration from the skin takes place, but this can be corrected by selecting foods with a lower water/energy ratio (e.g. banana versus watermelon, restricting cucumbers and tomatoes), and of course by not taking in fluid from sources other than food. Use of dried fruits can be made as a tradeoff between freshness and required fluid intake if necessary. Again, it helps if one has a sufficient understanding of what one is doing…]

(RW, CF) "(p26) and for meat eaters it increases the risk of eating toxins or pathogens that would be destroyed by cooking."
(DV) [Indeed, and that the various dangerous parasites among these pathogens do not have the same bad effect on the more naturally carnivorous animals that eat the same prey animals suggests that humans are not by evolution well suited to eating much of these animals, given that the evidence is actually more consistent with humans only having eaten much cooked food for no more than the most recent several ten thousand years, and not the 2 millions of years that Wrangham proposes, various raw processed plant foods being found more palatable, and therefore and for other (e.g. safety from other predators and from prey) reasons more eaten than raw meat.]

(RW, CF) "(p26) Recent studies indicate that low bone mass in the hips and backs of raw-foodists was caused by their raw diet."
(DV) [Maybe, by *their* raw diet, not by their *raw* diet, and more to the point:
Having examined the journal article (Fontana, Shew, Holloszy, Villareal; Low bone mass in subjects on a long-term raw vegetarian diet. Archives of Internal Medicine, 2005, 165; 684-689) it is quite clear that Wrangham has misrepresented their work – the vegetarians had a Body Mass Index (BMI) of 20 versus a BMI of 25 for the control group who ate a standard western diet (that is, the vegetarians had BMI 20% less than the control group), and the authors state that therefore the difference in bone density (about 15% less bone mineral density than the control group) is quite likely to be due to the difference in BMI, as BMI and bone mineral density are known to be physiologically related (with more weight to support proportional to body height, the bones get proportionately denser to enable the support of the proportionately greater weight. And also, as Fontana et al state "... bone quality also plays a role. It is therefore possible that raw food vegetarians with a low bone mass may not have an increased incidence of fractures, because of good bone quality."
This sort of misrepresentation is characteristic of a methodology of making up 'just so stories' to suit oneself – it clearly indicates poor science, and even on its own without that comprehensive

destruction of faulty work that I have amply provided in my review here, clearly indicates that the author's work in general is to be rightly viewed with great doubt…]

(DV) [(p26) Obviously enough I'd have thought, the association of raw food diets with low vitamin B12 is due to most raw food diets being vegetarian, not due to them being raw;
it's the HDL/LDL ratio that is mostly the point, not the HDL level by itself;
the randomized controlled trial(s) results are now in (Clarke et al 2011), and it is apparent that mildly elevated homocysteine levels as associated with vegetarianism (i.e. up to 30uM, the normal range being less than 10 or 12 or so uM, B12-unsupplemented long-term vegans being around 100uM, and the severe metabolic genetic disease levels being 200-500uM) are very unlikely to contribute anything significant to atherosclerosis…]

(RW, CF) "(p26) Maybe modern raw-foodists are so far removed from nutritional wisdom that they are just not choosing the right combinations of foods."
(DV) [Why leave the matter in doubt? Why not just refer to standard readily available food nutrient tables and find out what the pretty obviously right and wrong combinations by weight of foods to meet caloric needs are? It's really quite easy to do!...just average each of the nuts and seeds, the legumes, the cereal grains, the starchy root vegetables, the sweet fruit, and the leafy greens, to get the respective calories per weight, have a bit of a look and a think about it, then convert that data into weight of food group per some trial fraction of a daily energy requirement, have another bit of a look and a think about it – this is pretty easy stuff for anyone able to produce science worth presenting to the public!…]

(RW, CF) "(p29) Eskimo … … - in other words, smelly because they [(DV) stored fish] were partially rotten. Most people liked the strong taste. Jennes saw "a man take a bone from rotten caribou meat cached more than a year before, crack it open and eat the marrow with evident relish although it swarmed with maggots.""
(DV) [Too bad Wrangham couldn't see this as at least possibly

being what it almost certainly is – in an environment pretty obviously far removed from that humankind has evolved to live in, foods that are pretty far from what humankind has evolved to eat are eaten, and in the same way that humans can acquire the taste for cigarettes and strong alcohol, the first trial tastes of which are usually quite repugnant, come to develop a liking for the taste of them – surely this is telling us that what humans eat is very often not what we are physiologically and biochemically evolved to be best nourished by, but merely what allows us to survive and reproduce.
In nature, perfection of circumstance is rare, and the average circumstance is merely the main body of the Normal Distribution (Bell Curve) of something, far from perfect but enough to survive.
It is also possible that the maggot-eater may have been having a bit of a joke by way of generating disgust, as most young males have seen a friend do or done themselves in company at some time…
Also, note that Inuit ('Eskimos') also collect a variety of plant foods in summer and preserve them for use throughout winter.
Also, regarding p30 "… after a big kill … meat … people eat any number of times during the day, keeping their stomachs full to bursting, until all the meat is gone." (Which seems in any case inconsistent with the just previous detailing of the caching of excess meat for future use…), that consumption of more than 50% of calories as protein has been fairly conclusively demonstrated to be metabolically toxic (Bilsborough and Mann 2006) to the extent as being experienced as illness, such that fat must make up at least 50% of the source of calories in the case of near exclusive use of animal foods to provide not only protein for amino acids for building and repair, but also energy – animals have this sort of fat in cold climates, but most often don't in warmer climates near the equator, such as are our main sites of evolutionary background…]

(DV) [In Wrangham's very brief dealing with the eating habits of gatherer-hunters/ hunter-gatherers of the present-day and more recent historical era, where more detailed information is documented and available, he sums up:]
(RW, CF) "(p 31) Most fruits are preferred raw and are eaten in the bush, whereas roots, seeds and nuts are brought back to camp to be

cooked. Everywhere we look, home cooking is the norm. For most foods, eating raw appears to be a poor alternative demanded by circumstance."

(DV) [This is generally true, however Wrangham fails to note that gatherer-hunters/ hunter-gatherers of the present-day and more recent historical era generally live in the more marginal areas of the world, the better areas (particularly for growing grain) having been taken over by agriculture, so that the foods available, and their proportions, and their seasons of availability, are not the same as they would be in those better areas if they had not been taken over by agriculture. Furthermore, the use of fire to cook foods not only probably does increase the energy available from many foods cooked, but is more likely to be in practice because it enhances the flavor of many foods, and because the use of fire is a technological and quasi- or pseudo-magical status symbol, as well as offering warmth and protection from predators.

Moreover, more relevant to Wrangham's theory on the use of cooking having altered human physiology such as to decrease the size of the intestinal tract, there is no good or even suggestive evidence that the controlled use of fire has been the general case in human life for anything other than the most recent tens of thousands of years, far from the 2 million years that Wrangham posits, too short a time to have caused such an evolutionary change. Roebroeks and Villa (2011), as further detailed below in my review of Wrangham's Chapter 4 When Cooking Began, state that there is no evidence at all for the habitual use of fire before 300-400,000 years ago in Europe.]

(DV) [In Wrangham's dealing with "What happens to people who are forced to eat raw diets in wild habitats, such as lost explorers, castaways, or isolated adventurers simply trying to survive despite losing their ability to cook?", he fails to appropriately consider the relevance of the facts that: chimpanzees, living wild in their natural environment, where they are very knowledgeable about what food is available from their local area as the large majority of their lives is occupied in locating and eating the natural foods; and that for the gaining of that knowledge many primatologists accept that their relatively large brains (as for orangutans and gorillas, who have almost no, and little, respectively, use of the brain size for social interaction) are necessary, indicating that sourcing of their food is not an easy matter of just wandering freely about selecting from a cornucopia of readily available food, but rather a difficult matter requiring a great deal of local knowledge, good memory and memories of feeding related information, and some thinking/ problem solving; and that even with all this experience and knowledge of how to extract food from their environment, in most places they occupy, every year in the poor or off-season for their foods, food is in such short supply that juveniles who are not yet fully grown cease to grow.

If it is that difficult for chimpanzees to get sufficient food in their off-seasons, in the areas their whole lives have been immersed in for generations, how reasonable is it to think that a human, who does not even live wild off the land in their home place, could be suddenly landed in a strange new place and have the local knowledge and developed skills to be able to source enough raw food there to survive?

Not very reasonable, obviously enough if one thinks about the situation in the appropriate way.

That such people have not survived very well is no evidence against humans having evolved to be unable to satisfactorily meet their nutritional requirements from raw foods.

Wrangham's suggestion on p35 that nardoo seeds (in the case of the ill-fated Australian explorers Burke and Wills), or *any* seeds for that matter, ground into a flour but not cooked, eaten in as great a quantity as the modern human stomach and intestines could accommodate, would be unable to provide sufficient macronutrients (protein and energy (carbohydrate plus fat)) to meet human requirements is plainly ridiculous, as any glance at standard food nutrient tables, or other data in the food content scientific literature, will show – no competent nutritionist would contradict me here, as seeds are quite well known to be virtually universally very macronutrient-dense foods…]

(RW, CF) "(p36) Raw-foodists, it is clear, do not fare well."
(DV) [This is *not* correct when the diet is even *half*-sensibly composed of the different food groups! (see above)]
(RW, CF) "They thrive only in rich modern environments where they depend on eating exceptionally high-quality foods. Animals do not have the same constraints: they flourish on wild raw foods."
(DV) [Animals in the wild are far from always "flourishing" – much of the time they are hungry, and often enough in some state of starvation or semi-starvation – the world is generally most often not a utopia for the animals in it – survival is a struggle in many ways…]
(RW, CF) "The suspicion prompted by the shortcomings of the Evo Diet is correct, "
(DV) [No it isn't, because the Evo Diet Study was very poorly designed, and it's data very poorly collected and analyzed (see above).]
(RW, CF) " and the implication is clear: there is something odd about us. We are not like other animals. In most circumstances we need cooked food."
(DV) [Nonsense! See above, and see my own personal experiment of more than 20 years detailed immediately below:]

(DV) [(Written in 2010) I have myself carried out a personal *experiment* (*not* a belief system, please note, other than maybe a little very early on…) for more than 20 years (from 1986 to 2008) of eating only a raw vegan (exclusively plant food) diet, eating only 2 meals per day, late brunch and dinner, without any supplementation of vitamin B12 (and this, the lack of any vitamin B12, and very probably this alone, is the risky part of the experiment, and I do not advise being without some source of vitamin B12 as an ongoing thing, and I make explicit here that it would be unethical to subject a child to it).
To start taking vitamin B12 supplementation in 2008 was virtually the only change I have made to this, and this to reduce risk, because I had achieved informatively valuable endpoints (good health, with performance measures of it) already by that stage of the experiment.

I stake all the credibility that my qualifications in science can evidence on my claim that this is the absolute truth, with virtually no deviation or exception whatsoever, other than maybe 10 times or less a tiny scrap of cheese (to total a combined weight of certainly less than 10g, and very probably only a fraction of that), that was missed by my very thorough scrutiny of a salad made by someone else (rarely eaten), may have found its way into me.

Nor was there any oral sex other than less than half a dozen instances, to provide whatever small amount of vitamin B12 might come from that (and no, not due to lack of sex drive (the presence of females at university and elsewhere in fact inflamed my sex drive to an annoying and distracting height, with persistent erections and sexual thoughts intruding on study), rather due to focus on large workloads, restraint of the urge, and lack of fixed abode such as would better facilitate entering into an intimate life partnership).

In general vinegar flies were prevented from laying their eggs in my food supplies, the very large majority of which were not organically grown for budgetary reasons, and although some very very small amount of vitamin B12 probably was supplied by vinegar fly or other insect eggs and larvae, it would certainly have been orders of magnitude below that amount that apes get incidentally in nature from only their plant food, without even considering what they get from adult insects.

However, I have throughout eaten a large green salad with virtually every meal (only two per day, late brunch and dinner), and green leaves are the main source of cobalt atoms that are not already incorporated in vitamin B12 – therefore it is possible, indeed probable, that my large intestine bacteria did manufacture some vitamin B12 from that, which found its way down a concentration gradient in the moist raw-food-derived large intestinal contents, back up through the ileocecal valve and into the small intestine just on the other side of that valve, where vitamin B12 is absorbed by the human body.

My vitamin B12 levels for most of that 20-plus year experimental period were 40-100 picomolar (normal range 150-450 picomolar) (around 50 picomolar for at least the last 10 years of that period) and my homocysteine levels were 80-100uM (normal range less than 12uM).

Regards body muscle and nervous system health:

At the end of that period of 20-plus years of raw vegan diet unsupplemented by any vitamin B12, at the age of 45 years, I finished the 2007 Ironman Korea Triathlon (A 3.8 km swim followed by a 180 km bicycle ride followed by a 42km marathon run, the same distance as at the famous Hawaii Ironman Triathlon) at the 15th percentile, that is, with 15% of the field in front of me and 85% of the field behind me.
That is my whole of life best objectively measured athletic result, notwithstanding that it was the one most trained for (for three months of set regime coming off a substantial mix of add-lib running and cycling for a base) and tried hardest for. I have never manifested athletic champion genetic ability, and have carried since the age of 18 or so a hamstring tear with adhesions, and an only partly rehabilitated prolapsed intervertebral disk. I was in employment, though on less than the normal 30 (about 24) teaching hours per week, as an ESL English teacher throughout the time of the Ironman training.

Working on farms harvesting fruit in between contiguous years of university study, I have consistently averaged 6 full workdays per week, continuously for the 3 month periods, and my reputation as a hard and good worker on those many farms and amongst those I've worked alongside is good, better than average, though I have no hard proof of this (farmers, Mareeba, Nth. Queensland area: Howe, Nastase, Rogato, Morgante, Alba, Weaver, Bosnic, Di Stefano, Hartley, Del Ben, Mete, Iacutoni, Ravanello, Christensen, Cazzola). I always had nothing to eat and only water to drink till lunchtime at midday while working on these farms – lunchtime was usually one hour, but sometimes only half an hour – when lunchtime was only half an hour I cut my leafy green consumption to one mouthful at

each of the start and finish of the meal, and spent the rest of the short time eating peanuts and rolled oats (chewed dry and properly), and fresh raw mangoes (eaten more quickly than I like to (which is slowly), but not bolted) – my work performance in the afternoons was, like that in the mornings, above average.
Sometimes (say twice a week) I ran for 10-odd km after work, with a bicycle ride thrown in maybe once a week after work. Yes, the evening meal included a good gutful of mangoes, but the oily nut and seed intake was no more than 100-odd grams – and, somewhat annoyingly, it was usual for me to have more body fat (though not that much) at the end of the harvest season than at the start of it – mangoes are not as concentrated in energy as bananas, but they are at the high end of the range for fresh raw fruits…

And the old saying "Hunger is the best sauce" was well borne out at those farm work-day lunch hours, the raw peanuts and dry rolled oats tasting *incredibly* good, to an extent that the average westerner with their sense of taste deranged by food preparation and flavouring techniques, and their lack of proper physiological hunger due to chronic overeating relative to requirements, would be far from able to comprehend – and this is/was not due to me talking myself into it – as a psychology graduate I am fairly well armed against that – it was very real, and so marked as to be actually fairly surprising.

Since the start of the experimental raw vegan diet 20-plus years ago my resting heart rate has almost never been above 48 beats per minute, with the minimum ever a couple of 39 bpm amongst the 42bpm of my fitter early focal triathlon period in 1986 to 1991 – by most standards that indicates, in the absence of some pathological process, a pretty good level of fitness.
Only now at age 48 and a half years does it appear that I must concede some barely noticeable ageing, with my failure to go under 40 minutes for the 10 km 2010 Bridge to Brisbane fun run (40:55) on training that I think would have sufficed in 2008 (when I did 39:49), notwithstanding a little extra upper body muscle (pushups, chin-ups, dips…) and fat (1 kg ?), this compared to my personal best of 38:10 in 2001 and comparable half marathon personal best

of 1:27 in 2001 and wind(gale)-affected equivalent in 2006 of 1:28 (this would plausibly adjust back to 1:27 or less (for the howling gale), the personal best time I had trained fairly hard for), though at these times I had been doing more running and there was less (say 1-2kg) of upper body muscle and fat. Half marathons in 2010 of 1:34 for both of the only two.

There has throughout the experiment virtually always been mostly daily enthusiastic physical exercising, running and martial arts mostly, with a variable good serving of bicycling (road and MTB) thrown in when convenient.

Regards nervous system, i.e. brain, health (particularly relevant to low/absent extrinsic vitamin B12 intake):

In 1996, 10 years into the experimental raw vegan diet, I achieved approximately (the results distribution graph was very small and not lined so as to facilitate reading it very accurately) the 15th percentile (~15% of the others getting a better score) of the GAMSAT test that Australian university students must sit to gain entry to postgraduate medical schools to study medicine to be medical doctors – approximately 25% of those who sit the test are, after an interview assessment, admitted to the courses. Had I completed senior high school maths and physics before sitting that exam, which I hadn't, due to being an almost singularly unapplied and rudderless student in high school (as a junk-food eating omnivore, notwithstanding the lack of almost any steering committee…), I would almost certainly have achieved a score higher again than the 15th percentile.

(High school physics and Pure Maths were completed in 2000, 14 years into the experimental raw vegan diet, at the Queensland TAFE's Open Learning Centre, with the highest possible grade category of the three passing categories being achieved, with a bare minimum of tutor assistance).

In my most recently completed university degree, Master of Public Health, at one of Australia's most highly ranked universities, the

University of Queensland, in 2008, just after 22 years of the experimental diet, and only some months into vitamin B12 supplementation (which did not result in any detectable difference in subjective feeling, notwithstanding me being of course non-blinded to my conditions), my grade point average was 6, the possible maximum being 7 (though I strongly objected to the marking of work in one subject (graded 5) for which I had probably better qualifications for the assessable items than did the course tutor and coordinator, and in which it seemed my rightfully destructive criticism of the poor work we were required to write about was poorly received due its antipathy to the networking practices often employed by the incompetent in academia in securing each other's little niches in the avoidance of the dreaded manual labour…) though that was done on only a part-time basis while I carried out another research project, a monograph on the biomedical significance of homocysteine (which is raised in vitamin B12 or folate deficiency, among other conditions), which will soon be finished and uploaded to the Multifactor Health website, sections of it being there already.

And the state of my brain function may be further assessed by examination of this book here.
I have done most of my university degrees, commenced in 1991 after 5-odd years of raw vegan diet, at an overload of subjects (none repeated subjects), for example 50% overload in the third year of the BSc, 66% overload of PGDip PH&TM and second year psychology (this did have some, but not a full, content overlap in the statistics area), 75% overload (all subjects passed, at a D or 6/7 gpa, though I did have some prior knowledge) in one semester of Master of Medical Statistics and Master of Nutrition & Dietetics (which degree I was prevented from finishing, during the very *un*objectively assessed practical placements, the final semester of the course, by a most silly little coterie of liars and mental midgets (such as professor(!!...) Roger Hughes, very probably the stupidest PhD candidate that Prof Kerin O'Dea and Dr Shawn Somerset have ever nursed through that degree (I challenge them to deny it…), and the stupidest blockhead I've ever had as a Head of Department or lecturer anywhere in my university studies, who would not have

relished my adversarial and much better qualified and competent presence in the field of public health nutrition that he has elected to network his easiest possible to arrange of paychecks in…) such as are not in short supply in the dietetic profession, but that's another story and I'm sure the intelligent reader could surmise readily enough the general theme of my idiot-inflicted problems, due to my vegetarian"ism", there, for example by fellow student Allison Spiegel (previous 'leadership' course attendee somewhere…) who achieved (we corrected each other's exam paper) 6 out of 24 for a multiple choice clinical dietetics exam of 24 questions with 4 answers to choose from per question (The result you'd be most likely to get if you knew nothing *at all* about the examined material and selected answers totally randomly – my result was 22/24, the highest result in the class on that occasion…)).

Of course these academic challenges were far from easy for me – I submit that very few people would have found them easy… The regular daily physical exercise was continued alongside that.

Regards the eating/meal schedules:

Since 1986 or so I've virtually only ever eaten two meals a day, late brunch and dinner, with nothing other than water in between (and this water usually mostly in summer, the meals having their own generally nicely sufficient water content), excepting a couple of liters of fruit juice in the middle of the 6-odd hour Ironman training rides and the Ironman race (No-one else does these things without taking in some food along the way either, generally…) and a couple of marathon (42 km) runs and one 9 hour running traverse end-to-end of Korea's Baekunsan Provincial Park's mountain ridge system and a 6 hour run followed by 7 hour fastish walk to traverse the even more formidable Jirisan National Park end-to-end in one day (Yes, as a pure run, which was the intention, Jirisan beat me, but the one day full traverse just in itself is nevertheless quite a challenge – Hwaomsa Temple up to Nogodan along to Cheonwangbong peak (1,950m) down to Daewonsa Temple, if anyone's interested, and I'd advise not calculating your estimated time of completion solely on the basis of the map distance…).

I do spend some substantial time eating my food, say a total of 3 hours a day (but almost nil time in preparation and washing up…), but I am a slow eater, by choice, because I enjoy what I am eating, and do not like to rush my meals, and I have seen many others eat the foods I linger over in a mere fraction of the time I spend.

Even so, far from always feeling hungry, often enough I choose lower-calorie foods to prevent my tendency to overeat (which I have had since a child, though it has never lead to obesity as formally defined, only to, say, 5 kg of excess fat, mostly for some periods of childhood and adolescence) from adding unwanted and physiologically unnecessary fat to my carcass – for example, although I like bananas well enough, I usually reserve them for dinners before a race or long bicycle ride (more than two hours) or run (more than one hour).
I have only very, very infrequently eaten dried fruit in the 20-plus years of the experimental diet (not at all for most of the first 15 years – I did fall for the dried persimmons in Korea in 2000, and again in 2006-7 and 2009-10, but not heavily), mostly lately in an effort to be a bit more relaxed about life, but I have stopped this again now (for the present, I will have some more dried persimmons, hopefully, and Chinese dates, in the future…) as only a small amount of dried food at the end of a meal was adding unwanted fat to my carcass, and was reducing a little my appreciation of raw fresh fruit, which I usually find delicious, not merely tasty.

Furthermore I generally do not eat more than 100-200 grams of nuts and oily seeds per day, as to do so would be a gross excess of protein intake, and also bring my energy intake to an excess – and I can crack brazil nuts, almonds, macadamia nuts etc, with naturally shaped stones, faster than I can eat them – so obviously Professor Wrangham's (of Harvard University…) statements here (This description of my personal dietary experiment originally appeared some months ago in similar critical reviewing of Wrangham's chapter in Peter Ungar (Ed)'s (2007) book The Evolution of the Human Diet The Known, the Unknown, and the Unknowable, in my own book "The Natural Human Diet? The Optimal Human

Diet? The known and the unknown, the possible and the probable - an analysis of the evidence" (available free at the Multifactor Health website at: http://sites.google.com/site/multifactorhealth, but surely better purchased in print…)) and in his book Catching Fire How Cooking Made us Human (2009), that humans cannot meet their caloric requirements from raw plant foods, are so far wrong as to be laughable, if the possible reinforcement and spreading of this gross ignorance were not a very serious matter regarding the public health.

In short, If I were to spend more than approximately one hour per day eating the more concentrated raw natural plant foods available to me, at the speed that I've seen many people eat them, I would become more and more overweight by the week… if I were to spend two hours per day doing that, I would become morbidly obese, despite my regular exercise, soon enough!...

An old saying: "It is better to remain silent and be thought a fool, than to speak and remove all doubt." …]

Detailed review of Catching Fire
Chapter 2 The Cook's Body (p 37-53)

(RW, CF) "(p38) Domestic animals such as calves, lambs and piglets grow faster when their food is cooked,"
(DV) [Maybe, but maybe this depends on what food it is they're being fed, much of which bears little relation to the foods they and their wild relatives would naturally eat – for example, in the arguably fairly naturally disgusting intensive animal farming industries, when slaughter remnants and otherwise dead animals are ground up and used to feed animals that would otherwise not eat such matter…]
(RW, CF) "(p39) The spontaneous benefits of cooked food explain why domesticated pets easily become fat"
(DV) [This is plainly inadequate as, obviously enough, the biggest factors in pets becoming fat is over-feeding and under-exercise…]

(RW, CF) "(p40-42) Evolutionary benefits of adapting to cooked food are evident from comparing human digestive systems with those of chimpanzees and other apes. The main differences all involve humans having relatively small features. We have small mouths, weak jaws, small teeth, small stomachs, small colons, and small guts overall. In the past, the unusual size of these body parts has mostly been attributed to the evolutionary effects of our eating meat, but the design of the human digestive system is better explained as an adaptation to eating cooked food than it is to eating raw meat. … … "
(DV) [Firstly, Wrangham has failed to include here that human tooth enamel is thicker and stronger than that of any of the apes (Teaford 2007 (1)), and that this feature is possibly related to chewing more seeds, and cracking more nuts, seeds and nuts being generally among the most energy-dense of foods, and also to eating foods that have come from nearer or in the ground and so have more abrasive dirt particles in them. Furthermore that due to the movement of the fulcrum points of the jaw during evolution we actually do not have weak jaws (Daegling and Grine 2007) - the jaw musculature, though, is indeed reduced.
Secondly, Wrangham has failed, and it is a crucial failure that is

never addressed in Wrangham's theorizing in this book or Wrangham's chapter in Ungar et al (2007), to present the fairly well-known alternatives, in particular, that the increased consumption of seeds (oily and cereal) and nuts, generally among the most energy-dense of foods, along with a continued consumption of legumes, and an increased consumption of starchy root vegetables, along with the processing of these foods by cracking, grinding, mashing, peeling, drying and soaking using naturally formed stone tools, is at least as likely as the eating of cooked food to have resulted in those changes to the human anatomy (Lucas 2007).

Thirdly, Wrangham fails to note that chimpanzees' larger mouths are used not only for eating, but also very much as a weapon, them having also elongated canine teeth which are mostly used (in either biting or display) as weapons for fighting in defense, competition or offense, and to a lesser extent in killing the small amount of prey that they mostly only eat when they are very hungry in the off-seasons for their preferred plant foods, and that only constitutes less than 5% of calories, and then only for some (mostly adult males) chimpanzees in a group (Stanford 1998, despite his subsequent exaggeration and sensationalization of chimpanzee carnivory in his book of 1999, Kortlandt 1984, Kortlandt 1984 citing Kortlandt 1967, Teleki 1981 citing van Lawick-Goodall 1968, Yamagiwa et al 1996), the hunger so severe that growing juveniles stop growing, that is, there is some degree of starvation (Hladik 1981). As our evolutionary ancestors developed, fairly early on, relatively formidable weaponization they no longer needed the mouth for a weapon very much – the display of a large mouth and formidable teeth, or even their appearance in a face-to-face meeting or confrontation, might even have provoked a clubbing or a spearing, such as to provide one factor for such a mouth to be selected against in evolution.

Fourthly, on p41 Wrangham fails to note the possibility that the use of large lips in the fruit eating of chimpanzees and bats is quite likely not to be a feature that is a functional part of the ingestion of fruit in itself, but a feature for the removal of fruit from the reach of others who would if they physically could steal the food away from the possessor – use of a club extends one's reach and facilitates the timely warning and walloping of food thieves approaching with hands ready to grab one's food. The cooperation and sharing that were necessary for the use of food collection, transport and storing that are part of the social behaviour of humans in a group also reduce the need for such use of the lips in the eating of fruit. The absence of such a feature should certainly not be taken as an indication of a suitedness to little fruit eating, even though almost certainly humans are evolved to have replaced some of the great focus on fruit that apes have with a focus on starchy root vegetables. Humans are still very probably evolutionarily adapted to eating just as much fruit as starchy root vegetables, and for these to constitute most of the dietary energy intake (Vance 2010/11).]

(RW, CF) "(p42-43) The reduction in tooth size produces a well-adapted system: physical anthropologist Peter Lucas has calculated that the size of a tooth needed to make a crack in a cooked potato is 56 percent to 82 percent smaller than needed for a raw potato."
(DV) [This is plainly nonsensical or irrelevant as human teeth of the present size are ample for the biting off and chewing of raw potato, which I frequently do myself.]

(RW, CF) "(p43) In humans the surface area of the stomach is less than one-third the size expected for a typical mammal of our body weight, and smaller than in 97percent of other primates. … … But the large intestine, or colon, is less than 60 percent of the mass that would be expected for a primate of our body weight."
(DV) [These attempts at quantification are very poor, and quite useless – a "typical" mammal, given the very wide range of mammals and their foods, is firstly non-existent, and secondly even if taken as some sort of average would be irrelevant – likewise, "other primates" also vary greatly in their foods, from near-extremes of folivory to frugivory, and off to the side of this continuum into significant carnivory – in any case reference would need to be made to relative stomach size – the relevant comparison is that of humans with the *apes*; any other comparison is nearly useless unless we have the *human-ape* comparison available to start with…]

(RW, CF) "(p43) Great apes eat perhaps twice as much per day as we do because their foods are packed with indigestible fiber (around 30 percent by weight, compared to 5 percent to 10 percent or less in human diets). Thanks to the high caloric density of cooked food, we have modest needs that are adequately served by our small stomachs."

(DV) [Wrangham fails to note that most of the reason for this difference in weight of food eaten is that apes derive approximately all their water from the food that they eat – the only sensible comparison here would include adding the weight of what humans drink to the weight of what they eat.

Furthermore, a quick glance down any standard human food nutrient contents table will show that none of the fruits, leaf or root vegetables, cereal grains, or nuts come even close to being 30% "indigestible" fiber by weight – although some of our modern foods have been selectively bred to reduce fiber, this is unlikely to have had much effect on cereal grains, nuts or seeds, and the 30% seems likely to be an exaggeration – if by dry weight was meant, then it needs to have been stated, because that would be a very different matter, and would in any case bring us back to the question of the meeting of water requirements.

And again, the caloric density is just as likely to be due to a greater use being/ having been made of more energy-dense foods like nuts, oily seeds, starchy seeds including cereal grains, and the processing of them and starchy root vegetables and even some legumes and fruits by cracking, grinding, mashing, juicing, and drying – and it is very defective science on Wrangham's part not to note and properly explore this quite obvious alternative explanation that has been well enough noted in the literature (Lucas 2007, Sept 2007, Peters 207).

Furthermore, the causal association of various gastrointestinal diseases with insufficient dietary fiber has been pretty well established as a fact by now…]

(RW, CF) "(p44) Mouths and teeth do not need to be large to chew soft, high-density food, and a reduction in the size of jaw muscles may help us produce the low forces appropriate to eating a cooked diet."
(DV) [Actually, the changed jaw position has mostly compensated for the decrease in jaw muscle size, to retain high chewing forces (Daegling and Grine chapter 6 in Ungar (Ed) 2007). And again I mention that human tooth enamel is thicker and stronger than that of any of the apes (Teaford 2007 (1)).]

(RW, CF) "(p44) ... cooked diet. ... reduce tooth damage and subsequent disease."
(DV) [On the contrary, cooking starchy foods has been well enough documented in the literature to increase (often greatly), not decrease, tooth decay (Franck 1959, Price 1939, Cutress 2003, Mobley 2003, Gaqne 1993)]

(RW, CF) "(p46) ... our mouths, teeth and jaws are clearly not well adapted to eating meat unless it has been cooked."
(DV) [And I agree. Nor is our biochemistry (see p178-83).]
(RW, CF) "Dogs tend to keep food in the stomach for 2 to 4 hours, and cats for 5 to 6 hours, before passing the food quickly through the small intestine. By contrast, humans resemble other primates in keeping food in our stomachs for a short time, generally 1 to 2 hours, and then passing it slowly through the small intestine. Lacking the carnivore system of retaining food for many hours in our stomachs, we humans are inefficient at processing chunks of raw meat"
(DV) [And I agree, though the chewing of meat first to break it into smaller pieces and thus enhance digestion needs to have been more fully explored, though this is nearly irrelevant in the face of the other anatomical, physiological and biochemical considerations...]

(RW, CF) "(p48) Because the maximum safe level of protein intake for humans is around 50 % of total calories, the rest must come from fat, such as blubber [or nuts and oily seeds!], or carbohydrates, such as in fruits and roots.."
(DV) [Yes, but why fail to mention the nuts and oily seeds?...]

(DV) [(p49) More reference to the Giessen study and the supposed inability of raw food to meet human requirements – see my comments above in my reviewing of CF Chapter 1.]

(RW, CF) "(p49) Hunter-gatherers [(DV) gatherer-hunters, more accurately…] living on raw food might sometimes have found plant foods of an exceptionally high caloric density, such as avocadoes, olives, or walnuts. But no modern habitats produce such foods in abundance all year. Perhaps a few lost places would have had highly productive natural orchards until they were replaced by agriculture, such as the fertile valleys of the Middle East. But occasional productive areas would not explain the wide geographical range of human ancestors across Africa, Europe, and Asia by 1.8 million years ago."
(DV) [The fatal defect in Wrangham's reasoning here is the failure to address the possible and very probable significant consumption of starchy tubers and cereal grains, and their processing via pulping, grinding, drying and soaking, and their storage.]

(RW, CF) "(p52) Of the scores of chimpanzee foods I have tasted, I could imagine filling my belly with only a very few species, such as wild raspberry – but alas, one rarely finds more than a handful of these delicious fruits at a time."
(DV) [I accept this to be significantly true, but the failure to mention figs, which are among the most favored foods of chimpanzees, and many of which (not only our modern cultivated varieties) are very palatable to humans whose sense of taste has not been deranged through consumption of manufactured foods with altered and exaggerated taste, along with the fact that chimpanzees for significant periods of the year do in fact find enough figs to eat a daily gutful, is a major failure in Wrangham's work here.
Also, we need to know to what extent Wrangham's sense of taste has been deranged through consumption of manufactured and cooked foods with altered and exaggerated taste including through the use of added flavorings, in order to assess the meaning of his failure to find chimpanzee foods palatable – my own experience is that my own sense of taste by my teens had been substantially deranged by consumption of manufactured, cooked, added-flavor-

enhanced foods, and was then rehabilitated by a process of exclusion of those foods, a process that required (at the beginning…) a substantial degree of motivation and 'self-discipline', with the eventual result that a wide range of raw foods, including raw soaked grains and legumes, eaten in a state of physiological hunger, are very palatable and very often more than just palatable, in fact delicious.]

(RW, CF) "(p53) By contrast, if we were adapted to a raw-meat diet we would expect to see evidence of resistance to the toxins produced by bacteria that live on meat. No such evidence is known."
(DV) [And I agree.]

enhanced foods, and was then rehabilitated by a process of exclusion of those foods – a process that required at the beginning a substantial degree of motivation and self-discipline, with the eventual result that a wide range of raw foods, including raw soaked grains and legumes, eaten in a state of physiological hunger, are very palatable and very often more than just palatable, in fact delicious.]

(RW: CF 3/(p53) By contrast, if we were adapted to a raw-meat diet, we would expect to see evidence of resistance to the toxins produced by bacteria that live on meat. No such evidence is known.
(DV) [And I agree.]

Detailed review of Catching Fire
Chapter 3 The Energy Theory of Cooking (p 55-81)

(RW, CF) "(p58-59) The percentage of cooked starch that has been digested by the time it reaches the end of the ileum [the ileal digestibility] is at least 95 % in oats, wheat, potatoes, plantains, bananas, cornflakes, white bread, and the typical European or American diet (a mixture of starchy foods, dairy products, and meat). A few foods have lower digestibility: starch in home-cooked kidney beans and flaked barley has an ileal digestibility of only around 84%.
Comparable measurements of the ileal digestibility of raw starch are much lower. Ileal digestibility is 71% for wheat starch, 51% for potatoes, and a measly 48% for raw starch in plantains and cooking bananas."
(DV) [It seems very likely that Wrangham is correct in stating that cooked starchy foods are generally better digested of their starch than uncooked starchy foods. However, what is badly lacking here, and very important, is the quantification of the effect on the digestibility of raw foods of thorough chewing of the raw food verses insufficient chewing of the raw food.
Without this information the lower ileal digestibility %s of the raw foods as given here by Wrangham are meaningless for practical purposes – the %s given are probably for unchewed or substantially under-chewed raw food. People who are used to eating cooked food are in the habit of not chewing their food much, and tend to greatly under-chew raw food when eating it.
Furthermore it should be born in mind that it has been fairly well evidenced that this effect of cooking in increasing the starch digestibility of starchy foods has also the effect of greatly increasing tooth decay, as the cooked starch apparently deposits on or between teeth better than the analogous raw food starch, and is then able to be taken up more quickly or readily by oral bacteria which then produce the acid that attacks the teeth – and in areas where the cooked starchy food is a major staple food, and there are no dentists, the tooth decay is quite severe. (Franck 1959, Price 1939)]

(RW, CF) "(p60-61) Within thirty minutes of a person eating a test meal of pure glucose, the concentration of glucose in his or her blood rises dramatically, before returning to base levels in just over an hour. The effect of eating cornstarch is almost identical as long as it is cooked. But following a meal of raw cornstarch, the value of blood glucose remains persistently low, peaking at less than a third of the value for cooked cornstarch."
(DV) [The crucial flaw in Wrangham's thinking here is that it is not the *peak* of the blood sugar curve that is more relevant here, but the *area under the* blood sugar *curve*, which is much more closely related to the total amount or proportion of glucose absorbed (though a slower absorption rate will produce a smaller area under the curve if glucose is being transported out of the blood and into other tissues while it is still coming in to the blood from the gastrointestinal tract) – comparisons should be of equal timings.
Furthermore, in general a slower absorption rate is more desirable than a faster absorption rate, as a more constant blood sugar level (and its related blood lipid and ketone levels) is produced.
Also, from an evolutionary point of view, it is doubtful that having rapidly rising and falling blood glucose (and lipid and ketone) levels would have been advantageous over having more gradually rising and falling blood glucose (and lipid and ketone) levels – it seems more likely that this would rather be disadvantageous, in the absence of some peculiar circumstantial factor being at play to make it advantageous.]

(RW, CF) "(p61) Animal protein has been almost as important as starch in diets throughout our evolution,"
(DV) [This is sheer conjecture! Most scientists working in this area of knowledge in fact disagree with this statement. As detailed elsewhere here and in my book "The Natural Human Diet? The Optimal Human Diet? The known and the unknown, the possible and the probable - an analysis of the evidence." (see the Multifactor Health website: http://sites.google.com/site/multifactorhealth, but surely better purchased in print…) the evidence is very much against this statement being correct. In particular see the nutritional biochemistry I detail in the last several pages (p 178-183) of my reviewing of Catching Fire's Chapter 4 below, which fairly

conclusively evidences that meat has been no more than a very, very small proportion of the human ancestral diet as directed the evolution of our present nutritional biochemistry (vitamin B12 has been/ is important, mostly from insects, animal *protein*, particularly other than insects, has not). The important point here is that it is very poor science to state as a fact something that is much in dispute, without so much as noting the dispute, let alone giving some detailing of the evidence for and against.
Furthermore, the extent to which any importance is nutritional rather than socio-cultural is much in question, and the question needs to be at least noted.]

(RW, CF) "(p63) Such care [(DV) in cooking eggs] suggests that the hunter-gatherers knew better than the musclemen [(DV) eating raw eggs]."
(DV) [Another example of utterly circular reasoning by Wrangham!...
And their cooking of eggs is probably due rather to taste and texture preferences, food warmth, and social technological/magical status-related use of fire, than to nutritional considerations...]

(RW, CF) "(p64-65) ... in the large intestine bacteria and protozoa digest the food proteins entirely for their own benefit. ... ...
When the eggs were cooked, the proportion of protein digested averaged 91% to 94 %. ... ... However, in the ileostomy patients, digestibility of raw eggs was measured at a meager 51%. It was a little higher, 65%, in the healthy volunteers whose protein digestion was estimated by the appearance of stable isotopes in the breath. The results showed that 35% to 49% of the ingested protein was leaving the small intestine undigested. Cooking increased the protein value of eggs by around 40%.
The Belgian scientists considered the reason for this dramatic effect on nutritional value and concluded that the major factor was denaturation of the food proteins, induced by heat. Denaturation occurs when the internal bonds of a protein weaken, causing the molecule to open up. ... ...
Heat is only one of several factors that promote denaturation. Three others are acidity, sodium chloride, and drying,"

(DV) [Of interest, and I make no argument against it.]

(RW, CF) "(p66) Hunter-gatherers have likewise been reported mixing acidic fruits with stored meats. The Tlingit of Alaska stuffed goat meat with blueberries and stored salmon spawn mashed with cooked huckleberries. Many other North American groups made pemmican by mixing dried and pounded meat with various kinds of berries, and Australian Aborigines mixed wild plums with the pounded bones and meat of kangaroo."
(DV) [Of interest, and I make no argument against its truth.]

(RW, CF) "(p69) The same principles held, said Beaumont, with respect to meat [as to plant matter]. … … If tender and finely divided, they are disposed of readily; if in large and solid masses, digestion is proportionally retarded."
(DV) [And I agree – generally raw food requires more chewing...]

(RW, CF) "(p70) Beaumont's discovery that soft and finely divided foods are more easily digested conforms to our preference for such items."
(DV) [Wouldn't have anything to do with an evolution for predominant frugivory (fruit eating), by any chance?...]

(RW, CF) "(p76) … and since faster or easier digestion demands less metabolic effort, softer food might lead to energy saved during digestion.
(DV) [But not necessarily to energy saved overall, as the cost of digestion does not include all of the costs of distribution (including transport, storage and retrieval) and metabolism (the process of breaking up the absorbed molecules and forming other molecules from their parts, and then dealing with those formed molecules in similar ways, and so on. And if nutrient intake into the bloodstream is rapid, there may well be an added cost to sequestering the nutrient load deeper into storage and then bringing it back out from that deeper storage, compared to less storage requirement for a nutrient load that is delivered to the bloodstream more slowly.]
(RW, CF) "The idea should make sense when you consider the greater liveliness you feel after eating a light meal compared to a

heavy one: the light meal demands less work from your intestines and therefore makes other kinds of physical exercise easy."

(DV) [Wrangham's attribution of the feeling of greater liveliness to their being less work done by the intestines is quite erroneous: the feeling of torpor after a heavy meal is in large majority a biochemical response to the large intake of nutrient, and only in very small part due to energy expended by the intestines in the digestion of it. If one runs for two hours and a feeling of tiredness results, this is due in significant part to expenditure of energy, which can be readily roughly quantified by the increased heart rate and breathing occurring during the running – the energy cost of digesting a large meal for the intestines is relatively small, as evidenced by the fact that heart rate and breathing go up only a little, much less than when physical exercise such as running is being done.]

(RW, CF) "This energy-saving principle has been beautifully shown in rats given soft food. … …"

(DV) [No, it hasn't, the beauty is entirely in Wrangham's mind, see above and below.]

(RW, CF) "(p77) The researchers concluded that the reason the softer diet led to obesity was simply that it was a little less costly to digest."

(DV) [And Wrangham, lacking the background in this area of science to successfully critically assess the actual meaning of their work and the understanding in their conclusions, accepts the researchers conclusion as they give it. In fact, rather than the reason the softer diet led to obesity being *simply* that it was a little less costly to digest, it is more likely to be due mostly to the softer diet being *more completely digested*. Had Wrangham the acumen in this area of science that he here in his book ventures into, he would have noted that the researchers should have collected the dung of the rats and assayed it to assess what the proportion of the diet digested was in each of the soft and hard food pellet groups, and that them not having done this, it was very poor science to conclude as they did – lacking the acumen, Wrangham fails to understand this crucial point, which is somewhat surprising given the opportunity he has had, through the process of what he wrote before this section in the book, to grasp this fairly basic concept of

nutritional science.
Furthermore, no assessment was made of any differences there might have been in the physical activity levels of the two rat groups. To start with, the soft pellets were in fact of larger volume per weight because they were "puffed up like a breakfast cereal" so the sensation of fullness (quite possibly acting to reduce exercise) after a meal would be greater in the soft pellet group. Furthermore, for the reasons I have elaborated on just a little above here, differences in the rate of loading of nutrients into (and out of, into tissues) the bloodstream may have occurred (for example via insulin response), and resulted in differences in activity levels, particularly if the rats had been conditioned to do a certain amount of chewing per meal by the fact of them all having been fed standard (harder) lab rat chow pellets before the experiment – the researchers made no mention of having anticipated this possibility and having assayed eating schedules and activity levels – poor science by them, and poor science by Wrangham not to have understood that all these aspects required examination, consideration, noting…
The researchers of course provided him with the faulty conclusion, but good science is, after all, not about making up 'just-so stories' as one goes along, or accepting other peoples 'just-so stories', but rather, about rigorously examining all aspects possible of a thing in search of unknowns and alternative explanations, no matter how damaging they may be to the concepts one has accepted or cherished. And this fundamental, but not fundamentalistic, basic characteristic of good science should not need to be voiced at the professorial level at this point in the history of science, but here it has needed to be voiced, and more than once…]

(DV) [(p78-79) Similarly, the Burmese python study, as also appearing in the TV documentary on the topic of Wrangham's theory as in his book under consideration here, does not measure the activity of the pythons to see what differences there may have been – in fact, the pythons are out of sight in opaque plastic boxes inside the sealed container which is part of measuring their oxygen consumption and carbon dioxide production.
Though the claim that cooking and grinding each reduced the cost of digestion by about 12%, and both together by about 23% seems

roughly plausible, it is poor science to uncritically accept these figures without mention of the possible alternative explanations that are fairly obvious to anyone with a professional level of acumen in nutritional science.
It may well be that previous studies had established that snakes in all of these feeding conditions did approximately no physical activity, but the matter requires to be addressed, and it has not been. In any case the claim that pythons "are an ideal species in which to test the effects of cooking on the cost of digestion" is far from established regards the extrapolability of the significance of experimental findings to humans, because snakes in small sealed opaque plastic boxes are evolutionarily very very different from humans. Their reactions, for example with respect to levels of physical exercise, might well be quite different to those of free-range humans in (roughly) analogous dietary conditions…]

(RW, CF) "(p79) It makes sense that we like foods that have been softened by cooking, just as we like them chopped up in a blender, ground in a mill, or pounded in a mortar. The unnaturally, atypically soft foods that compose the human diet have given our species an energetic edge, sparing us much of the hard work of digestion."
(DV) [Wouldn't have anything to do with, rather, an evolution for predominant frugivory, by any chance?... nor that in evolution, processing of foods by pulping, pounding, grinding, drying and soaking have been used from a much earlier time than fire, and are more important regards our physiological nutritional evolution?...]

(RW, CF) "(p80-81) Why then do we like cooked food today?"
(DV) [The taste habits of a lifetime of socialization and enculturation wouldn't have anything to do with it by any chance?... That Wrangham fails to address the acquisition of tastes in various arenas by habit inculcation, in particular in the arena of food taste preferences clearly indicates a gross ignorance on his part of a phenomenon that is fundamentally basic to understanding food taste preferences, and therefore fundamentally basic to nutritional science, a phenomenon that any graduate in nutritional science should be aware of, as well as anyone having passed the most basic

first year university general psychology subject…]
(RW, CF) "… … Consider foie gras, the liver of French geese that have been cruelly force-fed to make them especially fat. The fresh liver is soaked in milk, water, or port, marinated in Armagnac, seasoned, and finally baked. The result is so meltingly soft and tender that a single bite has been said to make a grown man cry."
(DV) [What tripe!..]
(RW, CF) "Our raw-food-eating ancestors never knew such joy."
(DV) [What nonsense! See various of my comments above in reviewing CF Chapter 1: And the old saying "Hunger is the best sauce" was well borne out at those farm work-day lunch hours, the raw peanuts and dry rolled oats tasting *incredibly* good, to an extent that the average westerner with their sense of taste deranged by food preparation and flavouring techniques, and their lack of proper physiological hunger due to chronic overeating relative to requirements, would be far from able to comprehend – and this is/was not due to me talking myself into it – as a psychology graduate I am fairly well armed against that – it was very real, and so marked as to be actually fairly surprising.]

(RW, CF) "(p81) When our ancestors first obtained extra calories by cooking their food, they and their descendants passed on more genes than others of their species who ate raw. The result was a new evolutionary opportunity."
(DV) [Wrangham has failed to understand the need for a consideration of the combination of:
taste preferences, natural and acquired;
fallback foods used in times or areas of shortage of more naturally evolutionarily preferred foods;
population generating areas, favorable areas from which population spills out of, and population sinks, unfavorable areas fed population from favorable areas, in which populations do more dying than generating; and,
the differences in the likelihood of finding early archeological remains from these respective areas, that is, that human archeological remains are more likely to be preserved and found in the unfavorable population sink areas, rather than the favorable population generating areas, due to dryness of climate.]

Detailed review of Catching Fire
Chapter 4 When Cooking Began (p 83-103)

(RW, CF) "(p83-84) Archeologists are divided about the origins of cooking. Some suggest that fire was not regularly used for cooking until the Upper Paleolithic, about forty thousand years ago, a time when people were so modern that they were creating cave art. Others favor much earlier times, half a million years ago or before. A common proposal lies between those extremes, advocated especially by physical anthropologist Loring Brace, who has noted that people definitely controlled fire by two hundred thousand years ago and argues that cooking started around the same time."
[(DV) *opined* is the correct word, not "noted", for what should be done, as is not a fact or even established to be very probable; it is an opinion…]
(DV) [The crucial things to understand here are that it is necessary, in addressing the central question dealt with by Wrangham in his book here, to differentiate *use of* fire, *control of* fire, and *ability to make* fire *on demand*, which Wrangham does not even note the necessity of, let alone do…
Furthermore, that a phenomenon occurred in some few places, at some point in time, is not an indication that it was occurring at all places and over all of the period of time of relevance or question.
It is necessary to know the proportion of known sites of a particular era that have evidence of fire (Roebroeks and Villa 2011). And this information needs to be adjusted for in analysis, or considered in the light of, the possibility (indeed probability) that discovery bias of archeological sites has occurred due to the finding of animal bones (which generally remain long after plant food refuse) having been the trigger for the discovery of the site.
That Wrangham does not seem to understand these essentials for the successful analysis of the evidence, such as he incompletely provides just below, is indicated by his not even noting the necessity for such knowledge, let alone attempting to carry out the elaboration of it…]
(RW, CF) "As the wide range of views shows [(DV) No, it doesn't "show" it, it *suggests* it, which is not the same thing…], the archeological evidence is not definitive. Archeology offers only one

safe conclusion; it does not tell us what we want to know.
But though we cannot solve the problem of when cooking began by relying on the faint traces of ancient fires, we can use biology instead. In the teeth and bones of our ancestors we find indirect evidence of changes in diet and the way it was processed."
(DV) [Unfortunately Wrangham has not addressed the fact that modern human teeth, like pre-human teeth, have a thicker and stronger layer of enamel than any of the apes (Teaford 2007 (1)). And, even less fortunately, has not at all addressed the biochemistry (crucially important to the question) that I deal with in the last pages of my reviewing of this Chapter, see below (p 178-183)]

(RW, CF) "(p84-85) In the most recent quarter of a million years, there is sparkling evidence of fire control and even occasionally of cooking, by our ancestors and our close relatives the Neanderthals. The most informative sites tend to be airy caves or rock shelters, many of them in Europe.
In Abri Pataud in France's Dordogne region, heat-cracked cobblestones from the late Aurignacian period around 40,000 years ago, show that people boiled water by dropping hot rocks in it.
At Abri Romani near Barcelona, a series of occupations dating back 76,000 years includes more than sixty hearths together with abundant charcoal, burnt bones, and casts of wooden objects possibly used for cooking.
More than 93,000 years ago in Vanguard Cave, Gibraltar, three separate episodes of burning can be distinguished in a single hearth. Neanderthals heated pinecones on these fires and broke them open with stones, much as contemporary hunter-gatherers have been recorded doing, to eat the seeds.
Our ancestors were using fire in the Middle East and Africa as well. In a cave at Klassies River Mouth, a coastal site in South Africa from 60,000 to 90,000 years ago, burnt shells and fish bones lie near family-size hearths that appear to have been used for weeks or months at a time.
Between 109,000 and 127,000 years ago in the Sodmein Cave of Egypt's Red Sea Mountains, people appear responsible for huge fires with three distinct superimposed ash layers and the burnt bones of an elephant.

Charred logs, together with charcoal, reddened areas, and carbonized grass stems and plants, date to 180,000 years ago at Kalambo Falls in Zambia.
Back to 250,000 years ago in Isreal's Hayonim Cave, there are abundant hearths and ash deposits up to 4 cm thick. … …
Such sites show that people have been controlling fire throughout the evolutionary life span of our species, *Homo sapiens*, which is considered to have originated about 200,000 years ago."
(DV) [This is both an uncomprehending and possibly misleading thing to say, as the human line split from the chimpanzee line approximately 6,000,000 years ago, and all of the period(s) of time from then to now is of relevance to addressing the questions of what foods modern humans are physiologically and biochemically evolved to eat, and what role cooking may have had in this evolution…]

(RW, CF) "(p85-88) An ancient fireplace at Beeches Pit archaeological site in England securely dated to 400,000 years ago lies on the gently sloping bank of an ancient pond. Eight hand axes attest to the presence of humans. Dark patches about one meter (3 feet) in diameter with reddened sediments at the margins show where burning occurred. Tails of ashlike material lead down from the fires towards the pond, … … The knapper [(DV) person making stone tools by striking off pieces of rock] abandoned it, and two flakes from the series fell inward and were burnt, indicating that the toolmaker apparently had been squatting next to a warming blaze.
Another 400,000-year-old site, at Schoningen in Germany, has yielded more than half a dozen superb throwing spears carved from spruce and pine, together with the remains of at least 22 horses that appear to have died at the same time as one another, apparently killed by humans. Cut marks show that people removed meat from the horses. At the same site were numerous pieces of burnt flint, four large reddened patches about one meter in diameter that appear to have been fireplaces, and some pieces of burnt wood including a shaped stick, also made from spruce, that had been charred at one end as if it had been used as a poker, or perhaps held over coals to cook strips of meat. … …

Farther south, however, fire-using is strongly attested at 790,000 years ago. In a well-dated site called Gesher Benot Ya'aqov, next to Isreal's Jordan River, hand axes and bones were first discovered in the 1930s, and in the 1990s, Naama Goren-Inbar found burnt seeds, wood, and flint. Olives, barley, and grapes were among the species of seeds found burned. The flint fragments were grouped in clusters, suggesting that they had fallen into campfires. Nira Alperson-Afil analyzed [(DV) she *analyzed*!!... Gee!... now I *am* convinced...] these dense concentrations. She concluded that the early humans who made these fires "had profound knowledge of fire-making, enabling them to make fire at will."
(DV) [We need to know why she concluded that the early humans who made these fires were able to make fire at will, instead of concluding that lightning started the/a fire(s), and that these people kept it/them going or transported it/them anew several/ however many times to their preferred living base and kept them going for a while at a time there, and that, preferring to knapp their tools in front of the fire, pieces of flint accumulated there. Note that Wrangham makes no description here of the actual remains of campfires themselves – if there had been such, then it needs to be quantified how the pieces of flint were distributed in relation to the remains of the campfires, and this according to size, if the implication is that they used sparks struck from flint to start the fires at will...]

(RW, CF) "Archaeological sites between 1,000,000 and 1,500,000 years old include burnt bones (at Swartkrans in South Africa), lumps of clay heated to high temperatures associated with campfires (Chesowanja, near Lake Baringo in Kenya), heated rocks in a hearthlike pattern (Gadeb in Ethiopia), or colored patches with appropriate plant phytoliths inside (Koobi Fora, Kenya).
But the meaning of such evidence as indicating human control of fire is disputed. Some archaeologists find it totally unconvincing, regarding natural processes such as lightning strikes as likely explanations for the apparent use of fire. Others accept the idea that humans controlled fire in the early days of *Homo erectus* as well established. Overall, these hints from the Lower Paleolithic tell us only that in each case the control of fire was a possibility, not a

certainty."

(DV) [And I repeat again for the benefit of some less professionally expert readers what I noted just above, because it is crucially important: The crucial things to understand here is that it is necessary, in addressing the central question dealt with by Wrangham in his book here, to differentiate *use of* fire, *control of* fire, and *ability to make* fire *on demand*, which Wrangham does not even note the necessity of, let alone attempt to do...

Furthermore, that a phenomenon occurred in some few places, at some point in time, is not an indication that it was occurring at all places and over all of the period of time of relevance or question.

It is necessary to know the proportion of known sites of a particular era that have evidence of fire (Roebroeks and Villa 2011). And this information needs to be adjusted for in analysis, or considered in the light of, the possibility (indeed probability) that discovery bias of archeological sites has occurred due to the finding of animal bones (which generally remain long after plant food refuse) having been the trigger for the discovery of the site.

That Wrangham does not seem to understand these essentials for the successful analysis of the evidence, such as he incompletely provides just below the following excerpt from Roebroeks and Villa (2011), is indicated by his not even noting the necessity for such knowledge, let alone attempting to carry out the elaboration of it...

Roebroeks and Villa (2011) (verbatim):

"Volcanic eruptions can be an important factor in driving the natural fire regime in volcanically active areas. Lightning strikes—with an average global flash rate of an estimated 44 ± 5 flashes per second (15)—or spontaneous combustion cause natural fires that can burn archeological camp sites. The remains of such grass and forest fires can become associated with archeological find materials. Hence, charcoal, charred bone material, and heated flints do not necessarily indicate anthropogenic fires (2). At open-air sites, direct evidence for human fires, such as charcoal, can be easily removed by natural processes, including erosion by water or by wind (16). Thus, the context of possible fire indicators is important. If burnt bones, heated artifacts, and charcoal occur in caves or enclosed sites where the deposits are demonstrably in situ and are not

reworked by slope wash or debris flow entering the cave, then they can be considered good indicators of anthropogenic fires. Much of the debate on the history of human control of fire relates to the problem of the correct interpretation of possible fire indicators, especially in the absence of modern excavation techniques and detailed studies of site formation processes. … …
However, surprisingly, evidence for use of fire in the Early and early Middle Pleistocene of Europe is extremely weak. Or, more exactly, it is nonexistent, until ~300–400 ka. … …
The pattern emerging is a clear as well as a surprising one: Where we would expect habitual use of fire in the Paleolithic, i.e., the northern latitudes, we do not see any clear traces of it at all until the second half of the Middle Pleistocene. Cave sequences spanning the later part of the Early Pleistocene and the earlier part of the Middle Pleistocene in Europe do not have convincing evidence of fire. The number and quality of these early sites are significant, and this absence of evidence cannot be ignored. The simplest explanation is that there was no habitual use of fire before *ca.* 300–400 ka and therefore that fire was not an essential component of the behavior of the first occupants of the northern latitudes of the Old World."]

(RW, CF) "(p88) Evidence of humans controlling fire is hard to recover from early times. Meat can be cooked easily without burning bones. Fires might have been small temporary affairs, leaving no trace within a few days of exposure to wind and rain.
Even now hunter-gatherers such as the Hadza, who live near the Serengeti National Park in northern Tanzania, may use a fire only once, and they often leave no bones or tools at the fire site, so archaeologists would not be able to infer human activity even if they could detect where burning had occurred."
(DV) [This applies more in the case of fire being relatively easy to make on demand, and less in the case of a burning stick having to be carried to every new fire site, in the latter case it being relatively more convenient to maintain a fire at some fixed and rain-sheltered site.]
(RW, CF) "The caves and shelters that preserve relatively recent evidence of fire use tend to be made of soft rock, such as limestone,

which erodes quickly, so the half-lives of caves average about a quarter of a million years, leaving increasingly few opportunities to find traces of fire use from earlier periods."
(DV) [This may be, and I include it here in the interests of scrupulous objectivity and fairness in dealing with Wrangham's book here, but it is not clear to me why early humans and protohumans would choose limestone caves over other sorts of caves (possibly they are inherently more common, and this differing according to area, which ought have been clarified and quantified by Wrangham).]

(RW, CF) "(p89) The inability of the archaeological evidence to tell when humans first controlled fire directs us to biology, where we find two vital clues. First, the fossil record presents a reasonably clear picture of the changes in human anatomy over the past 2,000,000. It tells us what were the major changes in our ancestors' anatomy, and when they happened. … …
Humans do not eat cooked food because we have the right kind of teeth and guts; rather, we have small teeth and short guts as a result of adapting to a cooked diet. Therefore, we can identify when cooking began by searching the fossil record."
(DV) [Another classic example of the circular reasoning that is just one of several crucial and, even taken individually, fatal flaws in Wrangham's theorizing in Catching Fire…]
(DV) [Another crucial and, even taken individually, fatal flaw in Wrangham's theorizing here is the failure to so much as mention, let alone give due consideration to in the manner that is fundamental to doing good science of this sort, the alternative explanation: that the processing of *raw* food by the use of naturally formed stone and wood tools to pulp, grind, crack, crush, separate, and peel, *raw* foods, in conjunction with drying, storage and soaking, and the selection of more calorically dense *raw* foods such as oily nuts and seeds, cereal grains, legumes, insects (both adult and larval), and starchy root vegetables, is at least as likely to explain the evolved characteristics of our mouths and guts (Lucas 2007, Sept 2007, Peters 207).
Particularly when the only very recent evolution of the supralaryngeal vocal tract (that part of the vocal tract that includes

the mouth with its teeth, tongue and lips) to the form that allows the production of the full range of spoken sounds that modern humans are capable of (Capasso, Michetti and D'Anastasio 2008, Donald 1991), and that clearly is not a trivial matter in the competition for survival of an animal as inherently social and strategic as humans, is considered.]

(RW, CF) "(p90) Across the different locations the apes [chimpanzees and other apes] responded similarly. No apes preferred any food raw. They ate sweet potatoes and apples with equal enthusiasm whether raw or cooked, but they preferred their carrots, potatoes, and meat to be cooked.
(DV) [The apples being sweet fruit, a natural staple and favorite food type of apes; the sweet potatoes (as root vegetables) not being an ape staple food, yet being found sweet enough to be palatable; the other foods not being a natural preferred food type of apes, but when softened to a consistency more closely resembling some of their favorite fruits (e.g. figs and other softer fruits) being found acceptable.
It is a crucial defect in Wrangham's theorizing in his book dealt with here that he does not address the meaning of the fact that human salivary amylase, an enzyme that commences the digestion of starch in the mouth, has become genetically triplicated in humans (on average, and though there is some differences between populations having had different historical diets, there is more difference within any such populations than among them) after our evolutionary divergence from the line of that gave rise to the modern chimpanzee, which like the other apes has only one gene coding for salivary amylase (Perry et al 2007). This evolutionary feature fairly clearly indicates a much greater involvement with eating starchy foods. And if the starchy foods important in our evolution had been mostly cooked, then why the need for the evolution (and retention, note…) of a triplication of salivary amylase?...]

(RW, CF) "(p90) The Tchimpounga chimpanzees were particularly informative because there was no record of them having eaten meat previously, yet they showed a strong preference for cooked meat

over raw meat."
(DV) [In other words, they didn't like the taste of raw meat... Any chance that this taste preference is because they are, like humans very probably remain through our evolution, an animal that by natural evolved preference does not favor and consume much meat other than in times of a seasonal or other deficiency of more naturally preferred foods, as a fallback food, when in a state of great hunger?...]
(DV) [Too bad there was no comparison of preferences across food types as well as across raw/cooked states, taking into account of course what these captive apes had been conditioned to prefer, if that was possible... and too bad Wrangham has apparently failed to appreciate the potential significance and need for that comparison(s)...]

(RW, CF) "(p91-92) In 2004 such abilities in the human brain were reported for the first time. A team led by psychologist Edmund Rolls found that when people had foods of a particular viscosity in their mouths, specific brain regions were activated. Those regions partly overlapped with regions of taste cortex that register sweetness. ... ..."
(DV) [Wouldn't have anything to do with an evolution for predominant frugivory (fruit eating), by any chance?...]

(RW, CF) "Our ancestors were surely prepared by their preexisting sensory and brain mechanisms to like cooked foods in the same way. A long delay between the first control of fire and the first eating of cooked foods is therefore deeply improbable."
(DV) [And I agree, however, and crucially, it is fairly obviously incorrect to assume, as Wrangham implicitly does here in his book, that the beginning of the use of cooking, in one or a few places, is the same thing as the general and prevailing use of cooking in most places.]

(RW, CF) "(p92-93) A long delay between the adoption of a major new diet and resulting changes in anatomy is also unlikely. Studies of Galapagos finches by Peter and Rosemary Grant ... ... Selection pressure was so intense that only 15% of birds survived and the species as a whole developed measurably larger beaks within a year. … … In fewer than 8,000 years, mainland boa constrictors that occupied new islands off Belize shifted their diets away from mammals and to birds, spent more time in trees, became more slender, lost a previous size difference between females and males, and were reduced to a fifth of their original body weight. According to evolutionary biologist Stephen Jay Gould, this rate of change is not unusual. Drawing from the fossil record, he suggested that 15,000 to 20,000 years may be about the average time one species takes to make the transition into another."

(DV) [I have no argument with this (nor much expertise in this area of science…), but it needs to be borne in mind that the average here, in the absence of any statement to the contrary, must include the data from millions of insect and other very small species with very short lifespans, the number of which such species is much, much greater than the number of larger species, so reducing the average. And again I mention Wrangham's failure here in his book to address the possibility that the only very recent evolution of the supralaryngeal vocal tract (that part of the vocal tract that includes the mouth with its teeth, tongue and lips) to the form that allows the production of the full range of spoken sounds that modern humans are capable of (Capasso, Michetti and D'Anastasio 2008, Donald 1991), and that clearly is not a trivial matter in the competition for survival of an animal as inherently social and strategic as humans, is, like the processing of *raw* food by the use of naturally formed stone and wood tools to pulp, grind, crack, crush, separate, and peel, *raw* foods, in conjunction with drying, storage and soaking, and the selection of more calorically dense *raw* foods such as oily nuts and seeds, cereal grains, legumes, insects (both adult and larval), and starchy root vegetables, particularly if taken together, at least as likely to explain the evolved characteristics of our mouths and guts as the use of cooked food (Lucas 2007, Sept 2007, Peters 2007).]

(RW, CF) "(p93-94) … such rapid rates of evolution are sharply inconsistent with some previous interpretations of the effects of cooking. Loring Brace suggested that the use of fire for softening meat began around 250,000 years ago, followed by a supposed drop in tooth size that began about 100,000 years ago. This would mean that for at least the first 150,000 years after cooking was adopted, human teeth showed no response. Because such a long delay before adapting to a major new influence does not fit the animal pattern, we can conclude that Brace's idea is wrong. The adaptive changes brought on by the adoption of cooking would surely have been rapid."

(DV) [Firstly, it is fairly obviously incorrect to assume, as Wrangham does here, that the beginning of the use of cooking, in one or a few places, is the same thing as the general and prevailing use of cooking in most places.

Secondly, it is fairly obviously incorrect to assume, as Wrangham does here, in the absence of any supporting evidence (and contrary to the general opinion of the scientists of this field, references given above) that meat was a substantial enough proportion of the diet to have potentially had the vaunted evolutionary effect through the use of the cooking of it.

Thirdly, the element of circularity in Wrangham's reasoning here should be apparent enough to the scient or astute reader…]

(RW, CF) "(p95-96) Effects of cooking include extra energy, softer food, fireside meals, a safer and more diverse set of food species, and a more predictable food supply during periods of scarcity. Cooking would therefore be expected to increase survival, especially of the vulnerable young. It should also have increased the range of edible foods, allowing extension into new biogeographical zones. The anatomical differences between a cooking and a pre-cooking ancestor should be at least as big as those between a chimpanzee and a gorilla. So whenever cooking was adopted, its effects should be easy to find. We can expect the origin of cooking to be signaled by large, rapid changes in human anatomy appropriate to a softer and more energy rich diet.

The search for such changes proves to be rather simple. Before 2,000,000 years ago, there is no suggestion for the control of fire. Since then there have been only three periods when our ancestors' evolution was fast and strong enough to justify changes in the species names. They are the times that produced *Homo erectus* (1.8 million years ago), *Homo heidelbergensis* (800,000 years ago), and *Homo sapiens* (200,000 years ago). These are therefore the only times when it is reasonable to infer that cooking could have been adopted."

(DV) [Wrangham's search here is simple because it is simplistic. He makes the crucial failing of failing to note the *fact* that at this point in time of the evolution of the knowledge pertaining to the central *questions* (incorrectly in Wrangham's case here transformed as often as not into *assumptions*) of Wrangham's book here, the processing of *raw* food by the use of naturally formed stone and wood tools to pulp, grind, crack, crush, separate, and peel, *raw* foods, in conjunction with drying, storage and soaking, and the selection of more calorically dense *raw* foods such as oily nuts and seeds, cereal grains, legumes, insects (both adult and larval), and starchy root vegetables (Lucas 2007, Sept 2007, Peters 2007), is at least as likely to explain the evolved characteristics of our mouths and guts as the use of cooked food (which may then have actually had no effect at all, if it followed later enough in time, which it seems to me it probably did).

This particularly if taken together with the only very recent evolution of the supralaryngeal vocal tract (that part of the vocal tract that includes the mouth with its teeth, tongue and lips) to the form that allows the production of the full range of spoken sounds that modern humans are capable of (Capasso, Michetti and D'Anastasio 2008, Donald 1991), and that clearly is not a trivial matter in the competition for survival of an animal as inherently social and strategic as humans.

And again, the element of circularity in Wrangham's reasoning here should be apparent enough to the scient or astute reader, and becomes even more obvious in the next page or two of his book as he proceeds to examine what time the use of cooked food is likely to have occurred, on the assumption that the changes in tooth, mouth and gut anatomy could only be due to the use of cooked food … as he reaches a conclusion that: …]

(RW, CF) "(p98) The reduction in tooth size, the signs of increased energy availability in larger brains and bodies, the indication of smaller guts, and the ability to exploit new kinds of habitat all support the idea that cooking was responsible for the evolution of *Homo erectus*."

(RW, CF) "(p98-99) Even the reduction in climbing ability fits the hypothesis that *Homo erectus* cooked. *Homo erectus* presumably climbed no better than modern humans do, unlike the agile habilines. This shift suggests that *Homo erectus* slept on the ground, a novel behaviour that would have depended on their controlling fire to provide light to see predators and scare them away. Primates hardly ever sleep on the ground. Smaller species sleep in tree holes, in hidden nests, on branches hanging over water, on cliff ledges, or in trees so tall that no predator is likely to reach them."

(DV) [Wrangham here fails to understand that humans still climb trees well enough to make nests in them as apes do;

fails to understand that with the appropriate use of sentries in a sufficiently large enough group sufficiently well armed with naturally shaped clubs and stone projectiles, fire is not necessary for survival while sleeping on the ground at a strategic location (i.e., under climbable trees, with more weapons cached in them, with cover and clear ground in appropriate locations in the field of potential operations) instead of in the trees;

and that, in a big enough group (and we (hominids) are in larger groups, because we are smaller and less well armed, than recent gatherer-hunters), which is much the point in considering the survival methods of our earliest hominid ancestors, yes, we rely on *multiple* sentries, to wake us appropriately, and then we defend as a group (with our weapons of throwing stones, clubs and pointed sticks/ primitive spears) such as to allow the mothers and children to get up the tree(s) first. Or we are sleeping on a barely-accessible ledge, as the gelada baboons do for safety.

Maybe on pitch-black nights we slept in the trees, or spelled sentries more often so they could be sharper in their vigilance, and posted more sentries, or had more people awake more often.

Fire could have helped survival of smaller groups in sleeping on the ground, but would not have been essential for larger groups such as our hominid ancestors must have operated in for survival on the ground in daytime also.

As Wrangham himself notes: …]

(RW, CF) "(p101) *Homo erectus* was as poorly defended a creature as we are, unable to sprint fast and dependent on weapons for any success in fighting. … … In the bush, people [gatherer-hunters in modern times] lie close to the fire and for most or all of the night someone is awake. … … In a 12-hour night with no light other than what the fire provides, there is no need to have a continuous 8-hour sleep. An informal system of guarding easily emerges that allows enough hours of sleep for all while ensuring the presence of an alert sentinel."

(RW, CF) "(p102-103) Two kinds of evidence thus point independently to the origin of *Homo erectus* as the time when cooking began.
First, anatomical changes related to diet, including the reduction in tooth size and in the flaring of the rib cage, were larger than at any other time in evolution, and they fit the theory that the nutritional quality of the diet improved and the food consumed was softer."
(DV) [I agree.]
(RW, CF) "Second, the loss of traits allowing efficient climbing marked a commitment to sleeping on the ground that is hard to explain without the control of fire."
(DV) [No it isn't hard to explain at all – it's very easy, see just above here]
(RW, CF) "The only alternative is the traditional theory that cooking was first practiced by beings that already looked like us – physically members of the genus *Homo*."
(DV) [And if this is the traditional theory, then the evidence and reasoning for it should have been rigorously laid out in appropriate detail, for Wrangham's science here to have been at all good science. The science is not good because the possibility, crucial to that traditional theory, that the processing of *raw* food by the use of naturally formed stone and wood tools to pulp, grind, crack, crush, separate, and peel, *raw* foods, in conjunction with drying, storage and soaking, and the selection of more calorically dense *raw* foods such as oily nuts and seeds, cereal grains, legumes, insects (both adult and larval), and starchy root vegetables, is as least as likely to explain the evolved characteristics of our mouths and guts as the use of cooked food, has not at all been properly, as in even the making of a serious attempt, let alone well, as in getting it right, dealt with.]
(RW, CF) "If this were true, by the time our ancestors adopted

cooking, *Homo erectus* had long ago adapted to a soft, easily chewed diet of high caloric density. But as we have seen, cold-processed techniques such as grinding and blending provide relatively poor energy even when carried out by raw-foodists with modern equipment."
(DV) [As detailed very amply above in my reviewing of Chapter 1, this last sentence is sheer nonsense and wishful thinking.]

(RW, CF) "For more than 2,500,000 years our ancestors have been cutting meat off animal bones, and the impact was huge. A diet that included raw meat as well as plant foods pushed our forebears out of the australopithecine rut, initiated the evolution of their larger brains, and probably inspired a series of food-processing innovations."
(DV) [This is far from being an accepted fact, and it seems to me that reviewing of the literature clearly indicates that most professional opinions are to the contrary (references given above) – that such meat consumption was a very minor activity, a fallback food eaten mostly in times of shortage of the preferred plant foods that were in fact preferred over raw and often decomposing/rotting meat, and even then mostly in geographical areas that were population sinks for our forebears (or their offspring that were genetically less our forebears than those hominids of the areas that were population generating areas, which is a crucial point to understanding what our genetic heritage might be…). And it is poor science not to have noted the lack of consensus on this and not to have presented the arguments for the alternative.
Consider for example that whereas in evolution our forebears ate such a high proportion of the plant foods that contain vitamin C, that, unlike most other animals, we have entirely lost the ability to synthesize our own vitamin C, most of us retain a very highly efficient ability to absorb, transport and store vitamin B12, which is only found in animal foods – if our forebears had eaten as much meat as Wrangham is suggesting, we would not have retained through evolution this extremely highly efficient ability – bear in mind that bad genetic mutations are always happening in (coming into) the human gene pool, somewhat if not entirely randomly, and are removed from the gene pool by the disadvantages that they

cause resulting in less successful reproduction on average by those members of the population who carry the bad mutations.
The following summary section (p264-266) (I've transferred the references from the detailed section) of "Chapter 3 Diet related biochemistry and mutations – indicators of evolutionary adaptations to and for diet", of my book "The Natural Human Diet? The Optimal Human Diet? The known and the unknown, the possible and the probable – an analysis of the evidence", freely available from the Multifactor Health website at http://sites.google.com/site/multifactorhealth (but surely better purchased in print...) provides the basic information of the sort that a scientist competent in the area of investigation that Wrangham has ventured into would have acquired, analyzed the meaning of, and presented in a book such as Wrangham's here – that he has not done so obviously raises the question of whether he was/is competent to venture into this area of investigation...:

3.3 The large majority of animals make their own vitamin C; humans no longer do, and so must get their vitamin C from their foods, indicating persistence of a genetic adaptation to a diet higher in vitamin C than that of most other animals. That is, higher in plant foods including fruits.

3.4 Vitamin B12 intestinal uptake, transport and storage is mostly very well conserved. Saturation of intestinal B12 uptake occurs at a low 1.5-2.0 micrograms/meal in the context of the body's ability to store 3000-5000 micrograms (Shils et al 2006, Watanabe 2007).
Even western diet-eating omnivorous humans have a low breast milk B12 (one tenth that of cow milk) (Hay et al 2008), and their newborn babies have a low liver B12 concentration relative to adults (from 1/3 to 1/20). These facts taken together clearly indicate that humans are genetically adapted to a low-vitamin B12 diet with a regular intake of very small amounts relative to the amount found in animal foods (Maclaren 1981). What the relatively large storage is for is unsure. Most probably the diet to which humans are evolutionarily suited is one with a periodic regular intake of incidentally and deliberately eaten insect matter, larvae and adults, with meat eating comprising very little of the diet.

3.5 Human dietary protein requirements are very low (20-25g purified amino acids/ 40g food protein / day / person) (Millward 2001).
Human breast milk protein/energy ratio is lower than that of all other food types (less than 1/10 that of meat), and much lower (only a fraction) than that of other mammals (Shils et al 2006, Oftedal 1981) – bear in mind that a growing infant requires a higher protein/energy ratio than any other age group of human.
Consideration of human water requirements clearly indicates that if enough fruit or root vegetables, and leafy vegetables, are eaten to meet water requirements, then most of the human protein requirements will be met from those sources, leaving only a small requirement for more concentrated protein sources like legumes, nuts, oily seeds, cereal grains, or animal foods, the first four of which would round out the energy requirement also in the amounts needed to round out the protein requirement, as they are comprised mostly of carbohydrate or fat, compared to meat (and especially lean meat, i.e., wild game other than of very cold climates), which is comprised mostly of water.
These facts clearly indicate that humans are genetically adapted to a diet very low in protein, which strongly argues against the possibility that we have become evolutionarily adapted to a diet with a high or even a moderate proportion of meat.

3.6 Generally, for humans more than 50% of dietary calories as protein is overtly toxic (from the nitrogenous waste resulting), and illness is felt within days (Bilsborough and Mann 2006).
The excess nitrogenous waste resulting from the breakdown of dietary protein for energy is diuretic – that is, it causes an increased excretion of water as urine (Silberbauer 1981). This is undesirable in habitats where water is in short supply, and maybe even in any case, as predators lurk preferentially near groundwater supplies.
High-protein diets for humans are reported to increase the cost of digestion by about 30% compared to high-fat diets (Veldhorst, Westerterp-Plantanga, Westerterp 2009).
Taken together, this information is evidence further suggesting that humans are not genetically adapted to a high meat diet, but rather to a predominantly fruit and vegetable diet.

Claims of meat being more energy-dense than plant foods are very obviously absurd: raw meat is about 70% water; raw legumes, nuts and oily seeds have very little water, the large majority of the weight being taken up by about 20% protein on average, and the another 70-odd% being fully fat or carbohydrate, so obviously raw nuts (600 kcal / 100 g) (or oily seeds (600 kcal / 100 g), and also cereal grains (380 kcal / 100 g dry) for that matter…) are more energy dense than raw meat (190 kcal / 100 g, lean; 230 kcal / 100 g, regular).

3.7 Human starch digestion has been enhanced in evolution by the triplication of salivary amylase after the last common ancestor of chimpanzees and humans (Walker 2007, cites Samuelson et al 1990). The foods most probably involved in this were starchy tubers and cereal grains.
This information is evidence suggesting that we are evolutionarily genetically adapted to a diet with higher complex carbohydrate (starch) content than that of our common ancestor with the chimpanzee.
Legumes also have a substantial amount of starch, but consumption of these has probably been relatively constant throughout relevant periods of evolution, as legumes have been reported to be a dietary mainstay for virtually all relevant non-human primates (see also chapter 2, section 2.5) (Ripley 1985).

3.8 Regarding the supposed human inefficiency in biosynthesis of taurine, as in cats (obligate carnivores), vegan breast milk taurine was reported to be 30 times the level of cow's milk-based infant formula, and to considerably overlap in distribution omnivorous human breast milk taurine (Stipanuk, p549 of Shils et al 2006, cites Rana and Sanders 1986), suggesting that humans are just as likely to be evolutionarily genetically adapted to be able to meet their taurine requirements on a very predominantly vegetarian diet as to not be able to.

3.9 Regarding the supposed human inefficiency in elongating and desaturating 18 carbon fatty acids to their product 20 and 22 carbon fatty acids, docosohexaenoic acid (DHA) and arachidonic acid

(AA) (similarly to obligate carnivores), comparisons of vegan breast milk with captive and wild apes and monkeys, with similar levels found in vegan breast milk and the milk of wild apes and monkeys (Milligan and Bazinet 2008), indicates otherwise – that our relatively low fatty acid elongation and desaturation ability is evolutionarily genetically adapted to provide a sufficiency of the DHA and AA products on a very predominantly plant food/ vegetarian diet.

3.10 Preformed Vitamin A from animal food is far in excess of requirements, in western diets. The human ability to synthesize vitamin A from pre-vitamin beta-carotene from plant food sources has been retained through evolution. That humans have not lost the ability to synthesize sufficient vitamin A from the pre-vitamin A (beta carotene), commonly available from many different types of plant foods, despite high-meat diets leading to liver stores (relatively low at birth) of an average 450 milligrams, massively in excess of the 1 milligram Recommended Daily Allowance (McLaren 1981), is clearly consistent with human high-meat (assuming the liver would have been eaten) diets not having been part of our evolutionary history for long enough to have caused us to evolve to be suited for high-meat diets, and is consistent with humans being evolutionarily a predominantly vegetarian animal.

3.11 Adult lactase persistence in humans is almost certainly the result of only a few or several thousand years of milk consumption after weaning (Walker 2007). So, the persistent activation of the genes for lactase has become part of the genome in some small and generally geographically prescribed populations (i.e., in Europe) and their emigrants.

Clearly all other mammals on earth drink milk only as infants, then go on to eat the natural (if not always the optimal) foods of their species. The fact that a small number of human groups have evolved by a genetically degenerative (DNA *de*generation, which happens more readily than generation of new function) process to be able to digest the milk sugar lactose is not at all evidence that dairy products are the best source of calcium for humans of any age, as other nutrients, and the combinations of foods in meals and diets, and even other factors, need to be addressed in deciding what is the best source of calcium.

People should be skeptical of assumptions that our genome is "adapted for a Paleolithic lifestyle" (Cordain, chapter 19 [2007]; Eaton, chapter 20 [2007]), for not only were Paleolithic diets quite variable in time and place, but natural selection may occur more quickly than commonly assumed.

3.12 Fructose is a sugar found in fruit, and some vegetables (i.e., sugar beets) where it most often is present in an equal or lesser concentration that glucose.
Inborn errors of fructose metabolism (in Fructokinase, (Fruct)Aldolase B, Aldolase A and Fructose-1,6-diphosphatase) are very low in frequency (1/20,000 or less) (Steinman, Gitzelman and Van den Berghe in Scriver et al 2001, Homes 1993).
The very common 'fructose malabsorption' seems mostly due to the unnatural modern context of rapid ingestion of some high fructose:glucose ratio foods in the absence of other meal-balancing protein and fat foods, and to be otherwise beneficial to the digestive tract (Steinman, Gitzelman and Van den Berghe in Scriver et al 2001, Segal 2002).
Overall it seems that the human diet must have contained a large proportion of fruit throughout the periods of time that have produced the current rareness of inborn errors of fructose metabolism – the evidence here is for humans being by genetic evolutionary nature very much a fruit-eating animal. However, the evidence agrees with a moderate swing towards suitedness for starchy root vegetables, such that we might be suited for a diet of less fruit and more starchy root vegetables and cereal grains than apes are.

3.13 Celiac disease is a condition in which an allergic response to the cereal grain (wheat, barley, rye, oats, millet…) protein gluten causes structural and functional damage to the surface of the small intestine, causing reduced absorption of nutrients from digested food.
The prevalence of celiac disease, in those countries where reporting

can be relied on, is considered to be about 1% (Tack et al 2010, Boguerra et al 1998).
A 1% prevalence of celiac disease in the context of the predisposing infectious diseases being more prevalent in the higher population densities of the agricultural (and modern urban) era is hardly a very strong argument for humans not being evolutionarily suited to eating cereal grains – particularly as other foodstuffs are also the subject of allergies: peanuts and tree nuts around 1% each in a variety of countries, milk 0.4% in the USA, egg 0.2% in the USA, egg 2% in UK children, 0.1% in adults, shrimp 1% in the USA and shellfish 5% in an Asian country. These figures are only very roughly comparable indeed, as they need to be related to exposure frequency and schedule, and meal context, and corrected for socioeconomic confounders such as affordability, weaning, and related infectious and nutritional deficiency diseases – obviously cereal grains are almost certainly *the* most affordable of these and all foodstuffs, and are therefore very likely confounded in their association with allergy by a more unfavourable collection of socioeconomic confounders such as weaning, and relevant diseases.
Celiac disease's 1% prevalence is only a very weak evidence, or none at all, for humans not being evolutionarily suited to eating cereal grains, nor for cereal grains not having been a substantial part of the diet in various places in the human evolutionary past, as we can safely assume peanuts, tree nuts, eggs and shellfish were also.]

(RW, CF) "(p103) But according to the evolutionary evidence carried in our bodies, it would take the invention of cooking to convert habilines into *Homo erectus*, and launch the journey that has led without any major changes to the anatomy of modern humans."
(DV) [It should be obvious enough from the material presented immediately above, and Wrangham's failure to address any of it, that Wrangham's knowledge of the relevant evolutionary evidence(s) and its(their) meaning is far too poor for him to have a valid opinion on the matter.

To once again present that old saying: It is better to remain silent and be thought a fool, than to speak and remove all doubt…
For some, to repeat often enough is to come to believe…]

Detailed review of Catching Fire
Chapter 5 Brain Foods (p 105-127)

(RW, CF) "(p106) Attempts to explain the evolution of intelligence have sometimes appealed to rather specific advantages. Evolutionary biologist Richard Alexander argues that because humans practice warfare, and brainpower is critical for planning raids and winning battles, higher intellect could have been favored by a long evolutionary history of intense intergroup violence. But this hypothesis is undermined by chimpanzees, which behave in ways similar to warfare in small-scale human societies, but without humans' braininess. Violence between groups of chimpanzees is like a "shoot-on-sight" policy. Parties of males attack vulnerable rivals from adjacent groups whenever they encounter them, sometimes during incursions deep into the other group's territory in search of victims."
(DV) [This is plainly nonsensical – it is the stratagem(s) that differentiates human fighting from chimpanzee fighting – in particular the *intra*group social and interpersonal dynamics, in conjunction with the *inter*group dynamics. The similarity in chimpanzee and human is to a substantial extent only superficial, such as might appear to be the case to someone who had little knowledge of which such dynamics underlay or didn't the end result of one individual or group fighting with or killing another.]

(RW, CF) "Death rates from these interactions among chimpanzees are similar to those in small-scale societies of humans, yet chimpanzees are much less brainy than humans and only about as clever as their more peaceable relatives, bonobos, gorillas and orangutans."
(DV) [A larger brain and increased intelligence may indeed be required for different purposes in different species, even among close relatives such as the great apes, but a species that comes to have all of the purposes combined (e.g. humans), is likely to require some summed even greater brain size and increased intelligence than species that have only some lesser subset of all the purposes.]

(RW, CF) "(p106-107) Another suggested explanation for the evolution of intelligence is more ecological than social. This line of thinking proposes that intellect would be favored in species that occupy large home ranges, on the theory that wide-roaming creatures would need exceptional brainpower to mentally map their territories. And indeed, human hunter-gatherers cover huge areas compared to the ranges of apes and monkeys. But the correlation between range size and brain size does not generalize. Species of primates with larger brains are more intelligent, but they show no overall tendency to have larger ranges."
(DV) [That more or exclusively folivorous monkeys, eating mostly leaves, which are inherently more ubiquitously available than fruits, have smaller brains than more frugivorous monkeys, eating by preference and in some seasons mostly fruits, which are inherently less ubiquitously available and must therefore be searched for more widely and with a greater use of memory, has been fairly well established (Milton 2000, Clutton-Brock & Harvey 1990, Deacon 1990).
Which is in good agreement with the apes, which are all more frugivorous than folivorous (even though gorillas in mountain highlands have seasons of the year when fruits are only a small proportion of their diet).
And of which the orangutan, being a savagely solitary animal, seems a very good example of brain size and intelligence being at least in some cases very strongly related to having a diverse yet selective diet that requires the memory-facilitated and strategic coverage of a fairly large range in order to efficiently acquire the most favored of those diverse yet selected foods in greatest possible proportions.
Why fail to mention at least these basic relevant facts?... Convenience in expounding one's pet theory?...]

(RW, CF) "(p107) A more promising approach assumes that numerous kinds of benefits come from being more intelligent. Clever species can forage in a variety of creative ways, such as using grasses and twigs to extract insects from holes, or lifting stones as hammers to smash nuts.
(DV) [And how unfortunate it is that in Wrangham's book here the

variety of possible and probable processing(s) of *raw* food by the use of naturally formed stone and wood tools to pulp, grind, crack, crush, separate, and peel, *raw* foods, in conjunction with drying, storage and soaking, and the selection of more calorically dense *raw* foods such as oily nuts and seeds, cereal grains, legumes, insects (both adult and larval), and starchy root vegetables, and its relationship to intelligence, has not at all been properly, as in even the making of a serious attempt, let alone well, as in getting it right, dealt with…]

(RW, CF) "(p107-108) Big-brained species can also manage complex social relationships. Evolutionary psychologist Robin Dunbar found that primates with bigger brains or more neocortex live in larger groups, have a greater number of close social relationships, and use coalitions more effectively than those with smaller brains.
Brains pay off socially when they beat brawn. Relationships can change daily in primates that live in large groups, such as chimpanzees or baboons. Flexible coalitions in which two or more group members gang up on another member allow small or individually low-status animals to compete successfully for access to resources and mates. Coalitions are difficult to manage because individuals compete for the best allies, and an ally today may be a rival tomorrow. Individuals must constantly reassess one another's moods and strategies, and alter their own behaviour accordingly. Clever animals can be deceitful too, deliberately hiding their feelings by masking facial expressions, or screaming to pretend they have been attacked when their real motive is to rally supporters to chase a dominant individual away from food. The result is a soap opera of changing affections, alliances, and hostilities, and a constant pressure to outsmart others.
Most animals are not up to the cognitive challenges of juggling social alliances."
(DV) [Agreed, but mention of this must include noting the fact (and I do here again, for emphasis) that the orangutan, though a savagely solitary animal with minimal social interactions, seems a very good example of brain size and intelligence being at least in some cases very strongly related to having a diverse yet selective diet that

requires the memory-facilitated and strategic coverage of a fairly large range in order to efficiently acquire the most favored of those diverse yet selected foods in greatest possible proportions, and not so much related to social functioning.]

(RW, CF) "(p109) The social brain hypothesis does not explain these variations. It sets up this problem: if social intelligence is so important, why do some group-living species have smaller brains than others?
Diet provides a major part of the answer. In 1995 Leslie Aiello and Peter Wheeler proposed that the reason some animals have evolved big brains is that they have small guts, and small guts are made possible by a high-quality diet."
(DV) [The crucial flaw in Wrangham's reasoning here is that he has apparently failed to even realize the fairly obvious alternative: that a high-quality diet makes possible both big brains and a small gut – that the energy net from a large gut full of high-quality food is fairly obviously much greater than the energy net from a small gut full of high-quality food, much less address the possibility properly. Wrangham's reasoning on p110-114 is accordingly fallacious. This failure seems to be due to a scientific methodology more founded on making up 'just-so stories' as you go along to suit your pet theory, rather than a scientific methodology of objectively and rigorously analyzing the meaning of the relevant evidence…]

(RW, CF) "(p109-110) Aiello and Wheeler's head-spinning idea came from the realization that brains are exceptionally greedy for glucose – in other words, for energy."
(DV) [Firstly, as I evidence immediately above, the coming of the idea was by a process none to sensible, as Wrangham describes it here… the nature of the "head-spinning" might be of interest… Secondly, the brain is also able to use ketones for a large portion of its fuel and to function very well doing so.]

(RW, CF) "(p109-110) For an inactive person, every fifth meal is eaten solely to power the brain. Literally, our brains use around 20 % of our basal metabolic rate … … primates on average use about 13 percent of their basal metabolic rate on their brains, and

most other animals use less again, around 8% to 10%."
(DV) [In science, making the appropriate, meaningful comparisons is very often crucially important – in this case it is not the *average* primate that is the animal required for a meaningful comparison, but *apes*. Also, *in*active humans must be compared with *in*active others, and it is not clear from Wrangham's writing here that that is what was done.]

(RW, CF) "(p114) In actuality, that phase of our evolution occurred in two steps: first, the appearance of the habilines, and second, the appearance of *Homo erectus*. Meat eating and cooking account respectively for these two transitions, and therefore for their accompanying increases in brain size."
(DV) [This sort of rank conjecture, particularly, as it is in disagreement with the majority opinion, needs to be prefaced with something like "It seems probable to me that, on the evidence as I interpret it, ...", rather than be stated as an accepted fact, which it is not... if good science and not some arbitrarily assembled belief system dressed up as science is the aim...]

(RW, CF) "(p115) Chimpanzees have a cranial capacity of around 350 to 400 cubic centimeters. Australopithecines, with the same body weight as chimpanzees or even slightly less, had substantially larger cranial capacities, about 450 cubic centimeters."

(RW, CF) "(p115) During seasons of plenty, australopithecines would have eaten much the same diet as chimpanzees or baboons do when living in the kinds of woodland that australopithecines occupied – fruits, occasional honey, soft seeds, and other choice plant items. It was when fruits were scarce that australopithecines must have eaten better than their chimpanzee-like ancestors. Present-day chimpanzees that are short of fruit turn to items specific to their rain-forest homes, eating foliage such as the stems of giant herbs and the soft young leaves of forest trees. In their drier woodlands australopithecines would have found few such items. The most likely alternatives were starch-filled roots and other underground or underwater storage tissues of herbaceous plants. ... ... chimpanzees do dig for tubers at times with sticks,

and australopithecines would have been at least as well adapted: their chewing teeth are famously massive and somewhat piglike, suited to crushing roots and corms."
[Here Wrangham fails firstly to mention legumes, which are fairly well known to be a fundamental staple of primates in general, including apes (Ripley 1984); secondly fails to address the possibility of more nuts, seeds and grains being eaten also; thirdly fails to note that australopithecine chewing teeth are shaped very much like modern-day human chewing teeth, the difference being mostly in the size; and fourthly to elaborate on how australopithecine tool use, particularly in dealing with foods, might have been superior to that of chimpanzees.]

(RW, CF) "(p116) A dietary change from foliage to higher-quality roots is thus a plausible explanation for the first increase in brain size, from forest apes to australopithecines 5,000,000 to 7,000,000 years ago."
(DV) [This is fatally simplistic theorizing – the eating of starchy roots is just one part of the plurality of things potentially involved, maybe in different complements across time in different places – some of these things are the processing of food by the use of naturally formed stone and wood tools to pulp, grind, crack, crush, separate, and peel foods, in conjunction with drying, storage and soaking, and the selection of more calorically dense foods such as oily nuts and seeds, cereal grains, legumes, insects (both adult and larval), and starchy root vegetables.
And there are also social behaviour, language (signed or vocalized) use, weapons use and strategy, considered together as well as separately, in surviving on the ground further from the safety of trees and for longer periods of time, and in smaller groups…
Also, given that the number of skeletal remains of our early hominid ancestors is very small, it is not clear that the periods in time that the skull volume increased were not significantly longer than the sharp thresholds that Wrangham states them to be.]

(RW, CF) "(p116-117) During the second sharp increase, brain volume rose by about one-third, from the roughly 450 cubic centimeters of australopithecines to 612 cubic centimeters in habilines (based on measurements of 5 skulls). The body weights of australopithecines and habilines were about the same, so this was a substantial gain in relative brain size. Given the archaeological evidence, the big dietary change at this time was more meat eating, ”

(DV) [Lee and Wolproff (2003) presented (fig 2), on a graph of brain size on the y axis and time on the x axis, the data for all known (only 94 skulls or fragments of a skull) specimens of *Homo* from 50,000 to 1,800,000 years ago, excluding habilines. Aside from the questions that others have raised about the appropriateness of their statistical analysis (which suggested a continuous process (though exponential) of change in skull volume (brain size) over time, not a Punctuated Equilibrium process in accord with the sudden changes assumed in Wrangham's theories here in Catching Fire), just a *visual* inspection of the data in fig 2 makes it quite obvious that it is not possible that it may be stated as a fact (central to Wrangham's theorizing) that there were fairly sharply delineated points in time where the brain sizes of human evolutionary ancestors changed significantly. This applies even more so to the period of time before 1,800,000 years ago, from when there are very few skulls available, and spaced even further apart in time. There are quite plainly not enough skulls to enable even a reasonable, let alone a properly scientifically rigorous, conclusion that there were sudden increases in the brain sizes of human ancestors when Wrangham says there were....]

(RW, CF) " so meat should have made this brain growth possible."

(DV) [As neither the sharp increases in brain size during points or relatively narrow periods in time, nor the drastic increase in meat eating, are supported by the evidence (the evidence is quite substantially against both of these assumptions of Wrangham's), this step in reasoning is quite fallacious.]

(RW, CF) "To account for such a large increase in brain size, it seems likely that habilines processed their meat."

(DV) [And again, as neither the sharp increases in brain size during points or relatively narrow periods in time, nor the drastic increase in meat eating, are supported by the evidence (the evidence is quite substantially against both of these assumptions of Wrangham's), this step in reasoning is quite fallacious.]

(RW, CF) "Apes and humans are disadvantaged: their teeth cannot cut meat easily, their mouths are relatively small, and as William Beaumont noticed in the case of Alexis St. Martin, their stomachs do not process hunks of raw meat efficiently."
(DV) [Indeed, and if evolution can occur as quickly as Wrangham has noted, possibly correctly, that it can, and if there was such a drastic increase in the meat eating of human ancestors so long ago (2,000,000 years ago), then why didn't our dentition and nutritional biochemistry (see p 178-183 above) evolve in accordance with that? The answer is that the drastic increase in meat eating proposed (even more incorrectly, presented as an accepted fact…) here by Wrangham did *not* occur….]
(RW, CF) "Chimpanzees also show that eating unprocessed meat is difficult with ape jaws. They chew their animal prey intensely, but small bits of undigested meat sometimes appear in their feces."
(DV) [Likewise, because as through evolution meat has not constituted more than only a very small proportion of their diet(s), they haven't evolved much to be suited to meat eating…]

(RW, CF) "Perhaps because of this hard work and inefficiency, chimpanzees sometimes decline the opportunity to eat meat despite their usual enormous enthusiasm for it. … …"
(DV) [Far from having a "usual enormous enthusiasm" for meat, actually, in seasons and locations where their more preferred plant foods are available, chimpanzees *nearly always* decline the opportunity to eat meat – for example, Kortlandt 1984 citing Kortlandt 1967, and Teleki 1981 citing van Lawick-Goodall 1968, for reports of wild chimpanzees being very uninterested in either live prey or fresh carcasses placed alongside their trails, and other reports clearly indicate that chimpanzee carnivory occurs very much mostly in the seasons of scarcity of their more preferred plant foods (Yamagiwa et al 1996, Kortlandt 1984) , seasons when their preferred foods are so scarce that growing juveniles cease to grow, which could reasonably be defined as some degree of starvation (Hladik 1981).]

(RW, CF) "(p118) The habilines would have faced the same challenge. If they had relied on unprocessed meat for as much as half their calories "
(DV) [This is *extremely* unlikely, and most professional researchers and writers in this field of knowledge agree with me on this point – the archaeological evidence, and the large majority of the authoritative literature on it, indicates nothing of the sort (Potts 1992, Shea 2007; see my calculations on the Solutre data of Olsen 1989 in my review of Wrangham's Chapter 8 below (p 236-7); and note that Blumenschine and Pobiner 2007, Bunn 2007, Cordain 2007, Richards 2002, and Eaton, Shostak and Konner 1988 all plunge ahead into their assumptions of meat constituting a large proportion of our ancestral diet without so much as attempting the basic calculations from the evidence available of what that proportion might be), and moreover the nutritional biochemistry of humans as I have elaborated at the end (p 178-183) of my reviewing of Wrangham's Chapter 4 of Catching Fire above fairly conclusively indicates that meat eating has been a very, very minor proportion of our ancestral diet(s).]

(RW, CF) " and had eaten their meat as slowly as chimpanzees, with certain cuts of meat they would have had to spend several hours a day chewing it.
The digestive costs likewise would have been high, since the gut would have been busy digesting for many hours."
(DV) [If the digestive system would have been busy for as long as Wrangham has suggested it would be in the case of eating raw plant foods, then the digestive cost of digesting the raw meat, chewed only, would not have been any higher – the point is the incompleteness of digestion and the loss of an undigested fraction of the meat, chewed only, in the feces…]

(RW, CF) "(p119) Australopithecines probably used similar practices [(DV) adding mature tree leaves to chewed meat, purportedly to enhance reduction of the meat by chewing, by chimpanzees] when they caught gazelle fawns or other small animals.
Habilines had access to more advanced techniques. Their bones are

found close to stone hammers, fist sized spheres whose shapes provide vivid testimony of their repeated use. Habilines probably used the hammers partly to smash prey bones to extract the marrow. They also doubtlessly used the hammers to crack open nuts, as West African chimpanzees do, as well as to make other tools. In addition to these practices, stone hammers or wooden clubs could equally have been used for tenderizing meat. After habilines cut hunks of meat off the carcasses of game animals, they may have sliced them into steaks, laid them on flat stones, and pounded them with logs or rocks. Even relatively crude hammering would have reduced the costs of digestion by tenderizing the meat and breaking connective tissue. Because raw unprocessed meat is difficult to chew and digest, I suspect that this was one of the most important cultural innovations in human origins, enabling habilines to increase the nutritional benefit of meat and the speed with which they could eat and digest it."

(DV) [And again, this is fatally simplistic and selective theorizing – "their bones" implies a general state of affairs that may not at all reasonably be assumed, due to the very low number of habiline remains found (less than a dozen still in 2011 I assume), and the nature of biases in their finding (found in drier, less optimal areas more likely to be population sink areas than to be population source areas, and the findings sometimes being prompted by the tools being found first), and that there is the above-noted plurality of things potentially involved, maybe in different complements across time in different places – some of these things are the processing of food by the use of naturally formed stone and wood tools to pulp, grind, crack, crush, separate, and peel foods, in conjunction with drying, storage and soaking, and the selection of more calorically dense foods such as oily nuts and seeds, cereal grains, legumes, insects (both adult and larval), and starchy root vegetables.

And again, moreover, the nutritional biochemistry of humans as I have elaborated at the end (p 178-183) of my reviewing of Wrangham's Chapter 4 of Catching Fire above fairly conclusively indicates that meat eating has been a very, very minor proportion of our ancestral diet(s).]

(RW, CF) "Tenderizing meat would have reduced the costs of digestion by cutting the time that meat was in the stomach, and thus allowed habilines to divert energy towards their brains."
(DV) [See my comment immediately above, and again, the point (largely irrelevant with respect to meat eating and more relevant to consumption of plant foods in any case) is probably more the incompleteness of digestion and the loss of an undigested fraction of food, and less the efforts of the stomach before it passes its contents to the small intestine.]
(RW, CF) "Dietary shifts towards roots, meat eating, and meat processing thus can explain the growth in brains from a chimpanzee-like ancestor at 6,000,000 years to the habilines around 2,000,000 years ago."
(DV) [Ditto my three comments immediately above.]

(RW, CF) "(p 119-120) From then on, the increases in brain size were more continuous. The habilines cranial capacity of 612 cubic centimeters rose by over 40% to reach an average of 870 cubic cm in the earliest measured *Homo erectus*. The significance of this rise is complicated by a parallel growth in body weight, from the lowly 32-37 kg of habilines to a substantial 55-66 kg in *Homo erectus*."
(DV) [This is about a 70% increase in body weight, not a "parallel" increase…]
(RW, CF) "Unfortunately, body weights are hard to estimate from bones, and the number of specimens is small, so how much larger relative to body weights the brains of the first *Homo erectus* were than those of habilines, or whether they were relatively larger at all, is uncertain."
(DV) [This is fatally incompetent regards the making of appropriate quantitative comparisons – if cranial capacity rose by 40% while body weight rose by 70%, then very clearly brain size became smaller relative to body size, not larger. And, again, another example here of Wrangham's selective and biased method is that he rightly notes here, when it suits his pet theory, that due to the small number of specimens (see Henry and Wood 2007) any changes are uncertain (that is, probably or certainly not statistically significant), yet elsewhere, regarding earlier times when there were, as I've noted above, even fewer specimens, and

it does not suit his pet theory, he does not voice the same caveat.]

(RW, CF) "However, *Homo erectus* brains continued to increase in size after 1,800,000 years ago, averaging almost 950 cubic cm by 1,000,000 years ago.
Given the evidence and arguments I have offered that *Homo erectus* originated as cooks, the expensive tissue hypothesis suggests their eating cooked food caused their brains to grow. Once cooking began, gut size could fall and the gut would be less active, both trends reducing the cost of the digestive system."
(DV) [As what relevant evidence there is has been not even close to appropriately addressed by Wrangham in his book here, and as his arguments are therefore quite ill-founded and fallacious, whatever the expensive tissue hypothesis suggests, to whatever extent that it is correct in any case, it is only very weakly suggestive that it might be possible that the eating of cooked food had some probably if at all very small contribution to causing or allowing an increase in brain size.
And again, at risk of boring the reader with repetition, but the repetition being used so as to make my rebuttal commensurate in that regard with Wrangham's repetition of his theorizings in his book here, because, and again, for some, to repeat something often enough is to come to believe it, other things at least as likely to have been involved in causing or allowing an increase in brain size include the processing of food by the use of naturally formed stone and wood tools to pulp, grind, crack, crush, separate, and peel foods, in conjunction with drying, storage and soaking, and the selection of more calorically dense foods such as oily nuts and seeds, cereal grains, legumes, insects (both adult and larval), and starchy root vegetables.
And there are also social behaviour, language (signed or vocalized) use and the evolution of the human vocal tract, weapons use and strategy, population expansion and competition for resources, and intra-group and inter-group interactions considered together as well as separately, in surviving initially more on the ground further from the safety of trees and for longer periods of time, and in smaller groups, and later on the interactions becoming more sophisticated and complex.......]

(RW, CF) "(p120-121) The fourth notable increase in brain capacity occurred with the emergence of *Homo heidelbergensis* after 800,000 years ago. The increase was again substantial, leading to a brain occupying around 1,200 cubic cm."
(DV) [And again, just a visual inspection of Lee and Wolproff's (2003) presented (fig 2), on a graph of brain size on the y axis and time on the x axis, data for all known (only 94 skulls or fragments of a skull) specimens of *Homo* from 50,000 to 1,800,000 years ago, excluding habilines, makes it very clear that no sudden increase between 1,200,000 and 300,000 years ago is reasonable to even tentatively infer (there is *no* doubt that it would *not* be statistically significant) – if one was to take the scanty data as giving clear indications of changes at certain times, one would have to explain a *decrease* in cranial volume between 1,200,000 and 900,000 years ago from 1000 cubic cm to 800 cubic cm. This is a good, and I would hope fairly obvious to anyone who had passed a basic undergraduate first-year university statistics course, example of the number of specimens being far too scanty to enable even a tentative suggestion of cranial volumes changing sharply at the points or relatively brief periods in time that Wrangham has stated them to have. His method here is not even poor science, it is pure fictionalizing and the making up as one goes along of a 'just-so story' to suit one's pet theory.]
(RW, CF) "This was the impressive rise that Aiello and Wheeler attributed to the invention of cooking – mistakenly, I believe. It remains a mystery, inviting speculation."
(DV) [It doesn't remain, because it never was…]

(RW, CF) "More efficient hunting is a possibility. Hartmut Thieme's evidence of group hunting 400,000 years ago in Schoningen suggests a marked improvement in hunting skills over earlier eras."
(DV) [And again, a fatal lack of understanding of even the general concept of statistical significance (_1_ site, in Europe…), as well as of the biases in archeological site discovery, as well as of the relevant nutritional biochemistry (see p 178-183 above).]
(RW, CF) "This raises the possibility that meat intake, and perhaps therefore the use of animal fat, rose significantly before this time

and played a role in the evolution of *Homo erectus* into *Homo heidelbergensis*."
(DV) [It probably didn't rise significantly, because the evidence of our nutritional biochemistry (as I've elaborated at the end (p 178-183) of my reviewing of Catching Fire Chapter 4 above), among other evidence, is virtually conclusively against such a rise…]

(RW, CF) "Alternatively, cooking surely continued to affect brain evolution long after it was invented, because cooking methods improved. … …"
(DV) [An even basic competence in the research methodology applicable to this sort of observational study of a population(s) would dictate appropriate consideration of the reverse direction of causation – that brain evolution affected cooking methods…]

(RW, CF) "(p122-123) Such improvements in cooking efficiency could explain why there was a steady upward trend in brain size during the lifetimes of the early human species. Brains were notably bigger in late *Homo erectus* than in early *Homo erectus*, and in late *Homo heidelbergensis* than in early *Homo heidelbergensis*."
(DV) [This is sheer nonsense, as just a visual inspection of Lee and Wolproff's (2003) presented (fig 2), on a graph of brain size on the y axis and time on the x axis, data for all known (only 94 skulls or fragments of a skull) specimens of *Homo* from 50,000 to 1,800,000 years ago, excluding habilines, makes amply clear – it seems a plausible idea, but that is all that can be said, other than that a line of best fit placed through all of these four groups (but not both the pairs of subsets Wrangham names) would rise from left to right, probably statistically significantly, indicating an increase in brain size.]

(RW, CF) "Major breakthroughs such as meat eating and the invention of cooking cannot account for these smaller changes. The steady rise in brain size between the major jumps is most easily explained by a series of improvements in cooking techniques. Perhaps some particularly important advances enabled the prominent rise in brain size with *Homo heidelbergensis*."

(DV) [This is an obvious fixation by Wrangham on selected facts and even non-facts to support his theory, to the exclusion of other fairly obvious possibilities requiring enumeration, elaboration, investigation here – a plurality of things potentially involved, maybe in different complements across time in different places.
Some of these things are the processing of food by the use of naturally formed stone and wood tools to pulp, grind, crack, crush, separate, and peel foods, in conjunction with drying, storage and soaking, and the selection of more calorically dense foods such as oily nuts and seeds, cereal grains, legumes, insects (both adult and larval), and starchy root vegetables.
And there are also social behaviour, language (signed or vocalized) use and the evolution of the human vocal tract, weapons use and strategy, population expansion and competition for resources, and intra-group and inter-group interactions considered together as well as separately, in surviving initially more on the ground further from the safety of trees and for longer periods of time, and in smaller groups, and later on the interactions becoming more sophisticated and complex…]

(RW, CF) "(p123) The same possibility applies to the evolution of our own species, *Homo sapiens*, around 200,000 years ago. The gain in brain size was relatively minor, from 1,200 cubic cm in *Homo heidelbergensis* to around 1,400 cubic cm in *Homo sapiens*. Various modern behaviours are seen [(DV) evidenced, not "seen" – these are two very different things!] for the first time around this transition, such as the use of red ochre (presumably as a form of personal decoration), making tools out of bone, and long-distance trade. Increasing behavioral sophistication [(DV) very much the point!...] could also have happened in cooking techniques [(DV) very much less the point!...] … …
(p127) Although the breakthrough of using fire at all would have been the biggest culinary leap, the subsequent discovery of better ways to prepare the food would have led to continual increases in digestive efficiency, leaving more energy for brain growth. … …
Cooking was a great discovery not merely because it gave us better food, or even because it made us physically human. It did something even more important: it helped make our brains

uniquely large, providing a dull human body with a brilliant human mind."
(DV) [And again, for some, to repeat something often enough is to come to believe it….]

Detailed review of Catching Fire
Chapter 6 How Cooking Frees Men (p 129-146)

(RW, CF) "(p129) The Man-the-Hunter hypothesis has inspired such potent explanations of bonding between males and females that it has seemed to some researchers that no other explanation is necessary. In 1968 physical anthropologists Sherwood Washburn and Chet Lancaster wrote, "Our intellect, interests, emotions and basic social life, all are evolutionary adaptations of the hunting adaptation"."
(DV) [Wrangham fails to properly note here that the general consensus these days in 2010/11 and for some years and even decades previously, has been that the Man-the-Hunter hypothesis (Lee, de Vore 1968) was mostly fallacious (Panter-Brick 2002, Nestle 1999, Sept 2007, Potts 1992, Harding and Teleki 1981, Mann 1981 (cites also Jolly 1970, 1972, Isaac and Crader 1981), Hart and Sussman 2007 (though criticized by Muller 2007 for lack of thoroughness, though Muller himself fails to note the crucial lack of quantification and bias identification in archeology, and lack of examination of relevant nutritional biochemistry (see p 178-183 above) in contradictory/complementary work he refers to as overlooked or not addressed by Hart and Sussman while noting the ease of their "kicking around Ardrey and Dart" as early Man-the-Hunter theorists.)).
It is very poor science of Wrangham to have not properly noted the consensus, or lack thereof, and provided the counter-arguments and identifications of weaknesses.
Readers may refer to the contributions of various authors in Evolution of the Human Diet (Ungar, Editor, 2007), and my own "The Natural Human Diet? The Optimal Human Diet? The known and the unknown, the possible and the probable – an analysis of the evidence," also available free at the Multifactor Health website.]

(RW, CF) "(p130) In fact, I believe that cooking has made possible one of the most distinctive features of human society: the modern form of the sexual division of labor."
(DV) [It is fatally defective science of Wrangham's here to have not appropriately addressed the possibility that other food

(transport and) processing methods require at least equal preliminary consideration: carrying, grinding, mashing, peeling, crushing, cracking, drying, and soaking.
Also, and fairly obviously so, weapon development, manufacture and use in defense of body and territory.]

(RW, CF) "(p131) The Hadza are modern-day people. Neighboring farmers and pastoralists trade with them and marry some of their daughters. Government officials, tourists and researchers visit them. The Hadza use metal knives and money, wear cotton clothes, hunt with dogs, and occasionally trade for agricultural foods. Much has changed since the time, perhaps 2,000 years ago, when they last lived in an exclusive world of hunter-gatherers." … …
(DV) [Indeed, and to which it may be added that the lands they occupy are more marginal in their nature regards richness in plant foods, than the areas that were agriculturally developed, and therefore the observed dietary practices of the Hadza, like those of other gatherer-hunters/ hunter-gatherers of historically recent and present times, are likely to differ from the dietary practices of gatherer-hunters/ hunter-gatherers who lived on the lands that are now under agriculture, before the inception of agriculture.
Exposure to foreign diseases, also needs to be mentioned.]

(RW, CF) "(p132-133) They [(DV) The Hadza] stop occasionally to pick small fruits that they eat on the spot. … … The digging is hard and uncomfortable but it does not take long. A couple of hours later the women's karosses – cloaks made of animal skins – are covered in piles of thick, brown, foot-long roots. These ekwa tubers are a year-round staple for the Hadza, always easily found. … … She hopes the men will bring some meat to complete the meal. During the evening hours several men return. Some have honey, a few have nothing, and one arrives with the carcass of a warthog. … …
(p133-134) Tierra del Fuego's bitter climate provided few plant foods, so while men hunted sea mammals, women would dive for shellfish in the frigid shallows.
In the tropical islands of northern Australia, there was so much plant food that women brought enough to feed all the family and still found time to hunt occasional small animals. Men there did

little hunting, mostly playing politics instead. Although the specific food types varied from place to place, women always tended to provide the staples, whether roots, seeds, or shellfish. These foods normally needed processing, which could involve a lot of time and laborious work. Many Australian tribes prepared a kind of bread called damper from small seeds, such as from grasses. Women gathered the plants and heaped them so that their seeds would drop and collect in a pile. They threshed the seeds by trampling, pounding, or rubbing them in their hands, winnowed them in long bark dishes, and ground them into a paste. The result was occasionally eaten raw but was more often cooked on hot ashes. The whole process could take more than a day. Women worked hard at such tasks because their children and husbands relied on the staples women prepared.

Men, by contrast, tended to search for foods that were especially appreciated but could not be found easily or predictably."

(DV) [The point is, the men escaped drudgery, even though they bore more combat injury.]

(RW, CF) "The aborigines continually craved meat, and any man was apt to declare, 'me hungry alonga bingy,' though he had had a good meal of yams and damper a few minutes before."

(DV) [It is very possible that this was a form of psychological rationalization of their escape from drudgery – and also that it has been exaggerated by the quoted author Phyllis Kaberry – note that in the description just above of dietary life in the tropical islands of northern Australia there is little indication of a continual craving for meat…]

(RW, CF) "(p136) Worldwide across foraging groups, however, men probably supplied the bulk of the food calories more often than women did. This is particularly true in the high, colder latitudes where there are few edible plants, and hunting is the main way to get food."

(DV) [The points here are that, firstly this is probably not true of groups living in more favorable climates, and almost certainly not true of groups living thousands of years ago in areas that have since been taken over by agriculture, and secondly that "supply" of

calories needs to be differentiated with respect to acquisition and preparation, and thirdly that there is the time taken in clothing (and weapons and other tools) manufacture to consider.
It seems to me that the usual pattern traditionally is that men more or less force women to do more of the tedious or uncomfortable work, even though there may be natural differences between males and females in predisposition to different types of labor, related to risk-taking, need for adventure or excitement, and predisposition to physical combat …]

(RW, CF) "(p138) But if food was raw, the sexual division of labor is unworkable." … …
(DV) [This is another fatally faulty part of Wrangham's theorizing – if food is eaten raw, there is still risky versus less risky gathering (or hunting), and defense during food processing (which is often characteristically noisy and repetitive enough to allow predators or enemies plenty of time to locate their target and choose the manner and timing of their attack approach), and defense of territories, including of stored food.]

(RW, CF) "The difficulty lies in the large amount of time it takes to eat raw food. Great apes allow us to estimate it. Simply because they are big – 30kg and more – they need a lot of food and a lot of time to chew."
(DV) [This last sentence is quite incorrect – All animals spend much time in whatever their particular species' combination of food acquisition and ingestion is – hunters have to spend more time in acquisition, herbivores have to spend more time in ingestion – this is after all perhaps the major population-limiting factor, amongst other often significant factors, predation, disease and intraspecific lethal combat – obviously enough I'd have thought…
In fact, smaller animals have generally higher metabolic rates, and so spend even more time in food acquisition – the voraciousness of the tiny insectivorous shrews has long been a fact of zoology well known to anyone with even a modest lay interest in animals, let alone biology graduates…]

(RW, CF) "(p139) Chimpanzees in Gombe National Park, Tanzania, spend more than 6 hours a day chewing. Six hours may seem high considering that most of their diet is ripe fruit. Bananas of grapefruits would slip down their throats easily, and for this reason chimpanzees readily raid the plantations of people living near their territories. But wild fruits are not nearly as rewarding as those domesticated fruits. The edible pulp of a forest fruit is often physically hard, and it may be protected by a skin, coat, or hairs that have to be removed. Most fruits have to be chewed for a long time before the pulp can be fully detached from the pieces of skin or seeds, and before the solid pieces are mashed enough to give up their valuable nutrients. Leaves, the next most important food for chimpanzees, are also tough and likewise take a long time to chew into smaller pieces for efficient digestion. The other great apes (bonobos, gorillas and orangutans) commit similarly long hours to chewing their food.
(p141) Wild monkeys spend almost twice as long chewing per day if their food is low-quality." [(DV) Low-calorie, he means, which is not necessarily the same thing – hamsters will always eat a substantial amount of green leaf vegetables even with excess sunflower seeds and water available, and I suspect apes do too…]

(RW, CF) "(p140-141) In the United States, children from 9 to 12 years of age spend a mere 10 % of their time eating, or just over an hour per 12-hour day. This is close to the daily chewing time for children recorded by anthropologists in 12 subsistence societies around the world. … … , adults chewed for even less time than the children. … … One might object that the people in the subsistence societies were observed only from dawn to dusk. Since people often have a big meal after dark, … … However we look at the data, humans devote between 1/5 and 1/10 as much time to chewing as do the great apes."
(DV) [This seems roughly correct to me, but whether it is a good thing is another matter. Cooked starchy foods are fairly well agreed to cause more, and much more dental caries (tooth decay) than do the same foods eaten raw (Franck 1959, Price 1939, Cutress 2003, Mobley 2003, Gaqne 1993references). Then there is the issue of the constancy of blood levels of sugar and fat, which is less for the

cooked foods. And the psychosocial issues of quick meals. And the proportions of nutrients, as well as tooth decay issues, when the very low calorie leafy green vegetables are displaced from the diet in the interest of the speed of eating.]

(RW, CF) "(p142) Thanks to cooking, we save ourselves around 4 hours of chewing time per day."
(DV) [Plant breeding requires mention for this to be true – a selected sufficient if not optimal raw food diet can be eaten in only 2 hours per day. The cooking, however, does require some time, and this can be quite a considerable amount of time. Also, the processing prior to cooking can also take up some time.
Various forms of processing raw foods can also reduce chewing time by hours per day. Though not all of these are optimal for health, for example removing fibre in the making of fibreless fruit and vegetable juices.]

(RW, CF) "(p142) Consider chimpanzees, who hunt little and whose raw-food diet can be safely assumed to be similar to the diet of australopithecines."
(DV) [It is a very important consideration in the evolution of the human diet that australopithecines are quite likely to have eaten more root vegetables and grains than chimpanzees…]
(RW, CF) "At Ngogo, Uganda, chimpanzees hunt intensely compared to other chimpanzee populations, yet males still average less than 3 minutes per day hunting."
(DV) [Which is equivalent to stating that other chimpanzee populations hardly hunt at all. Note that elsewhere in Catching Fire Wrangham finds it convenient to his theorizing to exaggerate the tendency of chimpanzees to hunt far beyond what he states it to be here…]

(RW, CF) "(p143) Human hunters have lots of time and walk for hours in the search for prey.
(DV) [And in the avoidance of the drudgery of the acquisition of some plant food staples, in search of excitement, in search of the opportunity to establish their martial virtue by proxy, and in the maintenance of territorial boundaries…]

(RW, CF) "A recent review of 8 hunter-gatherer societies found that men hunted for between 1.8 and 8.2 hours daily. Hadza men were close to the average, spending more than 4 hours a day hunting – about 80 times as long as an Ngogo chimpanzee."

(RW, CF) "(p143) Almost all hunts by chimpanzees follow a chance encounter during such routine activities as patrolling their territorial boundaries."
(DV) [Strongly suggesting they are in reality far from avid meat-eaters, unless very hungry, as they are in many locations in the off-season of their preferred plant foods, such that growing adolescents actually cease to grow, which in human terms we would term some degree of starvation.]

(RW, CF) "(p144) Therefore, a 5-hour chewing requirement becomes an 8- or 9-hour commitment to feeding. Eat, rest, eat, rest, eat. An ancestor species that did not cook would presumably have experienced a similar rhythm."
(DV) [Not if it was processing foods with naturally-formed or very primitively made stone and other tools, in a simple step up from what wild chimpanzees have already been well established to do with naturally-formed tools. It is poor science not to mention and address this possibility….]

(RW, CF) "(p144) Washburn [(DV) of the Man-the-Hunter theory] and other anthropologists [(DV) some! Far from all now do (see refs p 7-8 above)…] have proposed that the human division of labor by sex was based on hunting. They suggest that on days when a male failed to find meat, honey or other prizes, a female could provide food to him. As we now see, this would not have been sufficient, because a returning male who had not eaten during the day would not have had enough time left in the evening to chew his plant-food calories."
(DV) [Not if it was processing (cracking, grinding, peeling, mashing, extracting, drying, soaking) foods with naturally-formed or very primitively made stone and other tools, in a simple step up, through gradual improvements, from what wild chimpanzees have already been well established to do with naturally-formed tools. It

is poor science not to mention and address this possibility....]
(RW, CF) "The same time constraints apply whether our precooking ancestor obtained his staple plant diet by his own labor or received it from a female."
(DV) [That the *same amount* of time constraints could not possibly apply whether our precooking ancestor obtained his staple plant diet by his own labor or received it from a female, should be quite obvious to anyone examining the question objectively instead making up a 'just-so story' as they go along – there is *obviously* a *great deal* of difference between 1 person doing the food acquisition and preparation for 2 people, and 2 people doing the food acquisition and preparation for 2 people]
(RW, CF) "A division of labor into hunting and gathering would not have afforded consumption of sufficient calories, as long as the food was consumed raw."
(DV) [There is may well be some truth in this, but it seems to me that Wrangham has failed to differentiate it correctly and understand the real detail of it – it may be that to some extent cooking allows males to shift a greater fraction of the domestic drudgery of staple food acquisition and preparation onto females, while the males escape the drudgery by going off on the essentially less food-productive but more exciting hunting, which also serves as a proxy display of martial virtue within and between groups of men, reducing injury through actual combat in the context of lethal weaponization, and also probably partly thereby serves as a way of establishing and maintaining territorial boundaries as well as the more obvious patrol feature of hunting does (though there is some patrol feature in gathering staples too).
The necessity or benefit of this probably changed as our ancestors became less and less vulnerable to predation and so increasingly expanded in population size, bringing increasing competition for increasingly relatively limited resources.]

Detailed Review of Catching Fire
Chapter 7 The Married Cook (p 147-177)

(RW, CF) "(p147-148) Why should a female cook for him?" … …
(DV) [Because she needs him for defense for herself and her children from predators and other males, and for defense of territory and stored foods.]
(RW, CF) "It leads to the uncomfortable idea that as a cultural norm, women cook for men because of patriarchy. Men use their communal power to consign women to domestic roles, even when women would prefer otherwise."

(RW, CF) "(p152) Relying on cooking creates foods that can be owned, given or stolen. Before cooking we ate more like chimpanzees, everyone for themselves."
(DV) [Cooking is not the only action that creates foods that can be owned, given or stolen! Obviously enough to anyone with a rudimentary enough understanding of the subject to be able to sensibly venture into discussion about it, is that other actions, namely the gathering of foods and their transport, in particular of foods that (unlike *some* fruits) require some other processing before being eaten, and/or are able to be stored – for example starchy root vegetables that might be washed and/or peeled and/or pulped and/or have their juice extracted; nuts that need the use of tools to crack them open; seeds (whether or not requiring threshing for separation from the plant stems/parts carrying them) that may then be ground or soaked or dried and then soaked later; and fruits that may be dried, whether or not later soaked or even ground.
It is fairly obvious, and the general consensus is, that substantial use of these actions in dealing with food very probably predated the general use of cooking by a very long time (references given above). These matters are so obvious, and so obviously of central importance to the formulation of theories such as Wrangham's here, that to fail to so much as mention, let alone properly address them, is extremely poor science, or rather, part, by omission, or making up a 'just-so story' as one goes along…]

(RW, CF) "(p153) For culinary historian Micheal Symons, cooking promoted cooperation through sharing, because the cook always distributes food. … …
The contrast between communal and solitary eating is particularly pronounced among hunter-gatherers, for whom cooking is a highly social act, unlike eating raw food."
(DV) [Wrangham fails to note here (and apparently to understand…) that foods (that constitute(d) a large fraction of the diet) that were once eaten raw, socially after gathering, transport, processing and storage, as detailed just above, came to be eaten cooked.
And, that the present-day and recent historical hunter-gatherers/gatherer-hunters whose dietary activities comprise those reported, are living in more marginal areas because the richer areas were taken over by agriculture, and therefore their dietary habits are not necessarily those of hunter-gatherers/gatherer-hunters who lived in the areas that agriculture has now taken over (references given above) – the more extreme and therefore obvious of these being the Inuit, who live where there is little plant food, even though even these people eat more plant food than is commonly thought.]

(RW, CF) "(p154) Relying on cooked food creates opportunities for cooperation, but just as important, it exposes cooks to being exploited."
(DV) [The same can be said, and needs to be said, about the gathering of foods and their transport, in particular of foods that (unlike *some* fruits) require some other processing before being eaten, and/or are able to be stored – for example starchy root vegetables that might be washed and/or peeled and/or pulped and/or have their juice extracted; nuts that need the use of tools to crack them open; seeds (whether or not requiring threshing for separation from the plant stems/parts carrying them) that may then be ground or soaked or dried and then soaked later; and fruits that may be dried, whether or not later soaked or even ground.]

(RW, CF) "(p154) Cooking takes time, so lone cooks cannot easily guard their wares from determined thieves such as hungry males without their own food. Pair-bonds solve the problem."
(DV) [The nature of the theft problem with regard to thieves from within the group, and with regard to thieves from outside the group, needs to be differentiated. And needs to be related to defense of the group's territory, because if husbands are away hunting, then obviously they are not immediately available for defense of women, children and food at that time…]

(RW, CF) "(p154) Having a husband ensures that a woman's gathered foods will not be taken by others; having a wife ensures the man will have an evening meal."
(DV) [The point is more that having a wife ensures that the man will have an evening meal with the minimum of effort – in effect, one particular man takes part of (and a sustainable part of) one particular woman's food, rather than any man taking part of (and possibly not a sustainable part of) any woman's food.
Also, it is a fatal defect in Wrangham's theorizing that he does not address the matter of males wanting regular sexual activity with females, in a sustainable way which is not dangerous nor consume large amounts of time and effort in search and negotiation, nor cause resentment in other males of the group.]

(RW, CF) "(p154-155) The many beneficial aspects of the household, such as provisioning by males, increases in labor efficiency, and creation of a social network for child-rearing, were additions consequent to solving the more basic problems: females needed male protection, specifically because of cooking."
(DV) [Again, it is a fatal defect in Wrangham's theorizing that he does not properly consider that females needed male protection, and *long* before the advent of cooking, for defense for herself and her children from predators and other males, and for defense of territory and gathered and stored foods, and foods in preparation as noted above. Also that males generally want regular sexual activity with females, in a sustainable way which is not dangerous, nor consume large amounts of time and effort in search and negotiation, nor cause resentment in other males of the group.]

(RW, CF) "(p155) … cooking is necessarily a long and conspicuous process. In the bush, the sight or smell of smoke reveals a cook's location at a long distance, allowing hungry individuals who have no food to easily locate cooks in action."
(DV) [I agree. But it needs to be noted that other methods of food preparation may be nearly as revealing of location – the *sounds* of food preparation methods such as cracking, grinding, pulping, pounding and threshing…]

(RW, CF) "(p155) The idea that cooking has influenced social relationships in this way is supported by the intense aversion to competition shown by hunter-gatherers eating their meals."
(DV) [And again, it is a gross defect in Wrangham's theorizing that he does not properly consider that this could and probably does apply just as much to uncooked food gathered and/or processed aforehand, involving carrying, grinding, mashing, peeling, crushing, cracking, drying, and soaking, all of which activities very probably *long* predated the use of fire on demand for regular cooking…]
(RW, CF) "(p157) Chimpanzees fight over any food that can be monopolized, but the contests are fiercest over meat, producing a fracas that can often be heard more than a kilometer away."
(DV) [Because meat is usually eaten in seasons or times of scarcity of the more preferred plant foods, scarcity to the point that growing juveniles cease to grow, which in human terms would be defined as some degree of starvation – in other words, they are desperately and ravenously hungry for anything at all that can be eaten and digested…]

(RW, CF) "(p159) Even humans steal readily in various circumstances, so our species is not inherently uncompetitive. … … People who have the chance to take from members of a different social network have few qualms about doing so."
(DV) [I agree.]

(RW, CF) "(p160-161) … a system that keeps the peace at mealtimes among hunter-gathers and other small-scale societies. The system consists of strong cultural norms. Married women must provide food to their husbands, and they must cook it themselves,

though other family members may help. Social anthropologists Jane Collier and Michelle Rosaldo surveyed small-scale societies worldwide. "In all cases," they found, "a woman is obliged to provide daily food for her family." That is why married men can count on an evening meal. As a result, they have little reason to take food from women who are not their wives." … …
(DV) [I agree.]

(RW, CF) "(p162) Despite their [Tiwi hunter-gatherer wives of northern Australia] economic independence and key role in their husbands' status, they were "as frequently and as brutally beaten by their husbands as wives in any other savage society."
(DV) [In general this seems true enough to me.]

(RW, CF) "(p163) Peace in the camp is further cemented by the principle that unless a husband gives his blessing, a wife could feed no other man except her close kin. … …"
(DV) [The possibility that this is related in some way to the promotion in part of some extent of exclusivity of sexual services needs to be considered.]

(RW, CF) "(p163) In Western society, we take the principle of ownership for granted. But among hunter-gatherers, this manifestation of private ownership is noteworthy because it lies in remarkable contrast to the obligatory sharing of men's food in particular, and more generally to a strong ethos of communitywide cooperation.
So however hard a man labors to produce food, in hunter-gatherer societies his rights to the food are a matter of communal decision."
(DV) [This arrangement seems to me to be due to the benefit for survival of forging a tightly bonded fighting unit of primarily men for territorial (and stored food within that territory) defense and acquisition.]

(RW, CF) "(p164) [(DV) Of Australia's Western Desert Aborigines] A bachelor or married man who approaches someone else's wife in search of food would be in flagrant breach of convention and an immediate cause of gossip, just as a woman would be if she gave

him any food. …… Under this system, an unmarried woman who offers food to a man is effectively flirting, if not offering betrothal. ……
(p165) Cofeeding is often the only marriage ceremony, such that if an unmarried pair are seen eating together, they are henceforth regarded as married. … … so others take the opportunity to tease the new couple with jokes equating food and sex. … …
The husbands role is important not so much for his physical presence, but because he represents a reliable conduit to the support of the community."
(DV) [These phenomena do not necessarily imply the primacy of food (sharing restriction) over sexual activity (sharing restriction), as food sharing may lead to sexual activity, and vice-versa – though certainly a man may do without sex for a few days without much discomfort, whereas the same cannot be generally said for food – though of course the food that a male could arrange for himself could be just as nutritionally and taste-wise satisfying, whereas masturbation would not be as satisfying as sex with a woman.]

(RW, CF) "(p165-166) Among the Bonerif, as among many hunter-gatherers, sexual intercourse is not tightly restricted to marriage. Wives are free to have sexual relations with several man at the same time, and may do so even when their husbands protest.
Furthermore, they get little food from their husbands." … …
(DV) [It needs to be differentiated whether wives have sexual relations with several men at the same time as a function of the characteristics of their own husbands relative to the characteristics of the other men they have sex with: age, martial prowess, and status (even if only tacitly established and acknowledged) to be considered at minimum…]

(RW, CF) "(p166) Having a husband means that when social conflict arises, a good wife has an advocate who is a member of the ultimate source of social control [(DV) the patriarchy, the men]."
(DV) [A 'bad' wife who has had sexual relations with men other than her own husband may also have plural advocates who are members of the ultimate source of social control, and whose role in this control may be a function of their age, martial prowess, and

status…]

(RW, CF) "(p166-167) A link to the communal authority is critical, because the ability of victims to deter a bully or a persistent pest depends on their being a legitimate member of the community. … … Capital punishment provides the sanction that most completely enforces hunter-gatherer adherence to social norms, and it is in men's hands. Thus by being married (or, if unmarried, by virtue of being a daughter), a woman is socially protected from losing any of her food. Having a husband or father who is a legitimate member of the group, she is effectively protected by him."
(DV) [Because, it seems to me, this arrangement has a benefit for survival, of forging a tightly bonded fighting unit of primarily men for territorial (and stored food within that territory) defense and acquisition.]
(RW, CF) "In theory, cultural norms that oblige a woman to feed her husband but no other men could have arisen from a societal goal other than to protect women's foods. Such norms might have arisen from a desire to avoid conflicts in general, or from a concern for reducing adultery in particular. But these alternative explanations are unconvincing because men needed their wives specifically to cook for them, rather than merely to behave in a way that promoted community civility in general."
(DV) [These alternative explanations are far from unconvincing:
Firstly, protection of the woman and her children from predators and others not of her group is required at many other junctures in daily life than cooking, including many activities dealing with food other than cooking – finding, gathering, transporting, grinding, cracking, peeling, pulping, juicing, drying, storing and soaking.
Secondly, the social requisites for forging a tightly bonded fighting unit of primarily men for territorial (and stored food within that territory) defense and acquisition are very important, and to couch a reference to them in terms like "merely" and "promoted community civility" is to display a profound ignorance of these very important aspects.]

(RW, CF) "(p167) When an Australian Aboriginal wife deserts her husband, wrote Phyllis Kaberry, he can easily replace her role as a sexual partner but he suffers because he has lost someone attending to his hearth."
(DV) [It is very probably incorrect to suggest that the man in this case can easily get the sustainable, low-risk, low-time-cost, low-effort-cost sex every second day/night or so of the week that was available or enforceable with/from his wife, from elsewhere. And fairly obviously so…]

(RW, CF) "(p168) [(DV) On Mbuti pygmies, by Colin Turnbull] Without a wife a man cannot hunt; he has no hearth; he has nobody to build his house, gather fruits and vegetables and cook for him."
(DV) [He cannot go off on the daily hunting adventure, in between male socializing, as much as he would like to, because he'll have to be spending some time on the less amusing building of house, gathering of fruit and vegetables, and cooking. And of course there is the terrible demoralization from the loss/lack of social status attendant loss/lack of a wife…]
(RW, CF) "Examples like these are so widespread that according to Jane Collier and Michelle Rosaldo, in small-scale societies all men have a strictly economic need for a wife and hearth." Men need their personal cooks because the guarantee of an evening meal frees them to spend the day doing what they want, and allows them to entertain other men."
(DV) [In other words, there is a *want*, not a "need"…]
(RW, CF) "They can find opportunities for sexual interactions more easily than they can find a food provider."
(DV) [And again, it is very probably incorrect to suggest that men can easily get the sustainable, low-risk, low-time-cost, low-effort-cost sex every second day/night or so of the week that is generally available or enforceable with/from a wife, from elsewhere. And fairly obviously so…]

(RW, CF) "(p168) Among the Inuit, where a woman contributed no food calories, her cooking and production of warm, dry hunting clothes were vital: a man cannot both hunt and cook."
(DV) [Fairly obviously, I'd have thought, if the man is hunting for

only one person instead of for two or more, then he will in fact have some time to cook…]

(RW, CF) "(p169) Oosterwal recorded a comparable reason for wife stealing in New Guinea, where a woman's domestic contribution was critical because of the sago meal she prepared. Men wanted to give feasts as large as possible, so they needed women to organize the food. This led them to conduct raids on neighboring groups to kidnap wives for sago production. Captured wives were put to work at once. Their sexual services were an added bonus."
(DV) [Men *wanted* women to *prepare* the food, mostly…
Also, there is the issue of intragroup status as a function of martial virtue to consider – the extra wife will be an ever-present and highly visible symbol of this;
Also, there is the issue of population growth, maintenance or decline and the concomitant territorial expansion, maintenance or contraction to consider;
Also, there is the issue of hostage taking as a disincentive to indiscriminate killing of unidentifiable targets during any nighttime attack to consider;
There is no good evidence or reasoning provided by Wrangham that would incline one to think that these other reasons, taken together with the sexual services/utility, would not constitute much more of the reason(s) for wife stealing in New Guinea than any "need" for extra women to "organize" food…]

(RW, CF) "(p169-170) … the importance of marriage for a man in small-scale societies was universal. Collier and Rosaldo explained that a married man has status because once he has a wife he need never ask for cooked food and he can invite others to his hearth."
(DV) [And conversely, men with status, for example due to martial virtue or its proxies, are more likely to get a wife, and this part of this feedback loop might in fact be the primary causal part…]

(RW, CF) "(p170) He is also likely to eat well because men typically eat before their wives and have the choice of the best food. In Micheal Symons's words, men "demand selfless generosity from

women."
(DV) [I agree. Though it is necessary to consider that when fighting occurs, men are probably on average more likely to die therein, their women then transferring to a new man – I'm not putting this forward as an ethical justification, just that it must be considered in the summing up of the exchange of labor (or units of QALY (Quality Adjusted Life Years)) between the sexes.
Of course there is death and disability incurred in childbirth to consider also.
And then there are possible evolutionarily-derived genetic differences between the sexes in predisposition to excitement and risk, versus sustainability and safety, and in dominance versus submission, such that there may be different biologically-determined valuing of lifestyle characteristics.
These considerations made, I am still far from being an apologist for the dominance arrangements of historical and modern men over women.]

(RW, CF) "(p171) Catherine Perles was right in saying that cooking ends individual self sufficiency."
(DV) [Individual self-sufficiency was ended long before the advent of cooking, rather, it is very likely *never to have been at all*, right from the start of the divergence of our line from that of the chimpanzees, because our half-sized, newly-fangless, still as always relatively clawless, relatively slowly running, ill-disguised, and only very primitively weaponized ancestors the australopithecines, much like the modern baboons in this respect, would almost certainly have needed to function as a group to survive in the more terrestrial mode of life that they were diverging into from the more arborial mode of life of their and the chimpanzee ancestor's common ancestor (which was almost certainly a social animal).
And in the evolution on from the australopithecines all the way through to modern humans, it is almost certain, for the reasons repetitively noted by me here above regarding the need for defence of women, children, territory, stored food, and food in preparation (and of individuals often noisily engaged in preparing it) that sustainable self sufficiency of individuals, or even nuclear family

groups was never generally the case.
And this should be fairly obvious to anyone looking objectively at the evidence of combat realities for such creatures, and not just making up 'just-so stories' as they go along…]
(RW, CF) "Cooking need not be a social activity, but a woman needs a man to guard her food, and she needs the community to back him up."
(DV) [I agree, but unfortunately this is a very inadequate addressing of the whole situation, and of the alternative and/or complementary explanations to various extents contradictory to Wrangham's theorizing here, that logic and good science at even a basic level would require be addressed.]

(RW, CF) "(p171) We do not know how linguistically skilled our ancestors were when cooking was adopted."
(DV) [Indeed, and it is a major failing in Wrangham's theorizing here that he does not address the fact that the human supra-laryngeal vocal tract only evolved to be able to make the full range of sounds that modern humans can as little as less than 400,000 years (Capasso, Michetti and D'Anastasio 2008) and maybe only 40,000 years (Donald 1991) ago. Nor how this might relate to the evolution of human jaw and tooth structure and function.]
(RW, CF) "Language is needed nowadays to enforce culturally understood rules, and because a woman's food is made secure by her being able to report on a thief's activity. But at least we can say that three of the key behavioral elements found in the hunter-gatherer system – male food guards, female food suppliers, and respect for each other's possessions – are found in other animals, suggesting that a primitive version of the modern food-protection regime could have evolved rapidly among early cooks."
(DV) [The suggestion is more sensibly that it is very likely that a primitive version of the modern food-protection regime evolved *long before* the advent of cooking, in relation to food finding and gathering, transport, processing with naturally-formed or primitively-formed tools (grinding, cracking, peeling, pulping, juicing, drying, and soaking).
Language enables the competitive improvement of such arrangements and of other related defensive and offensive inter-

and intra-group actions and interactions.
Fairly obviously so. And it is fairly obviously very poor science not to have stated these fairly obvious alternative possibilities and addressed their probabilities…]

(RW, CF) "(p174) A woman's need to have her food supply protected is unique among primates and provides a sensible explanation for the sexual division of labor.
The proposal that the human household originated in competition over food presents a challenge to conventional thinking because it holds economics as primary and sexual relations as secondary."
(DV) [This is not a challenge to conventional thinking – it is very much a part of conventional thinking – the only challenge by Wrangham here, and I posit a very unsuccessful one, is that he has not elaborated an even barely comprehensive collation of the parts of the economic mechanism that are relevant to the question. These parts, or most or many of them, are quite well known to professional workers in this knowledge field. I have elaborated enough of them here above to illustrate the fatal deficiencies in Wrangham's theorizing in his book here, such that my repetition, despite the benefit to some readers in providing repetition somewhat commensurate with Wrangham's own repetition, is becoming annoying…]

(RW, CF) "(p174) Anthropologists often see marriage as an exchange in which women get resources and men get a guarantee of paternity."
(DV) [Men, and our primitive ancestors in particular, were almost certainly more interested in having sex than in having guarantees of paternity! The reproductive biological facts of paternity seem very likely to me to have been mostly a mystery to the large majority of our pre-historical ancestors. It's nearly certain that apes do not consider paternity, they consider sexual activity – and surely australopithecines likewise – and probably *Homo erectus* likewise…
And surely most *paleo*anthropologists, and even most anthropologists in general in considering the stages of human prehistory that Wrangham is dealing with, would agree with me…]

(RW, CF) "In that view, sex is the basis of our mating system; economic considerations are an add-on."
(DV) [This is a misrepresentation of the common consensus including of what possible complement of economic mechanisms may have contributed to the evolution of the behavior over the relevant pre-historical time periods.]

(RW, CF) "(p174) But in support of the primary importance of food in determining mating arrangements, in animal species the mating system is adapted to the feeding system, rather than the other way around. A female chimpanzee needs the support of all the males in her community to aid her in defending a large feeding territory, so she does not bond with any particular male. A female gorilla, however, has no need for a defended food territory, so she is free to become a mate for a specific male."
(DV) [The female gorilla still requires defense of self and offspring from the largest predators, even if the number of potentially successful predators is much less than for chimpanzees due to the size difference.
Also, the definition of "free" is perhaps a little peculiar here – unlike in chimpanzees, the dominant male gorilla is brutally and totally dominant, brooking almost no detectable sexual competition whatsoever – the subordinate males would not dare, and she will cooperate with the dominant males sexual advances sufficiently (allowing for pregnancy and the availability of other females) or receive a beating, unless I'm mistaken…]

(RW, CF) "(p175) Food, it seems, routinely drives a man's marriage decision more than the need for a sexual partner.
Furthermore, food relationships appear to be more tightly regulated than sexual relationships. Among the Bonerif, husbands disapproved of their wives having sex with bachelors, but the bachelors did it anyway. Husbands were relatively tolerant of their wives having sex with other husbands, perhaps because promiscuous sex involved less threat of losing her economic services than did promiscuous feeding."
(DV) [It is very poor science to select one case and put it forward as representative of the general situation – the Bonerif are one

group of people among hundreds or thousands. What is required is some multi-group survey with some determination of what both the average and the variance is…
And again, it needs to be differentiated whether wives have sexual relations with men other than their husbands as a function of the characteristics of their own husbands relative to the characteristics of the other men they have sex with: age, martial prowess, and status (even if only tacitly established and acknowledged) to be considered at minimum…
And again, part of their method may well be that all men get some sufficient minimum of sex, just as they share food, in order that all the men in a group will be bonded together tightly enough to fight as a committed group in the warfare (necessary for territorial maintenance or acquisition, when population density is at the carrying capacity of the land, given that infanticide (in the absence of sufficient predation or disease to contain the population) was not always viewed as a more desirable alternative to combat) that due to weaponization is more lethal than combat generally is between individuals of a species.]

(RW, CF) "(p176-177) … but among hunter-gatherers who are similarly egalitarian, husbands are liable to beat wives if the evening meal is merely late or poorly cooked.
When there is a conflict, most women have no choice: they have to cook, because cultural rules, ultimately enforced by men for their own benefit, demand it."
(DV) [This seems generally correct to me.]

(RW, CF) "(p177) The idea that cooking led to our pair-bonds suggests a worldwide irony."
(DV) [Cooking almost certainly didn't lead to our pair bonds.]

(RW, CF) "(p177) Cooking brought huge nutritional benefits."
(DV) [The benefits may well not have been huge in comparison with other food preparation methods. It may at one phase of evolution have allowed an increase in the human population through the consumption of foods that were otherwise absolutely inedible due to toxins or extremely laborious of processing due to

toughness – but the compromise of health, for example the great increase in dental decay caused by cooking starchy foods instead of eating them raw (even if otherwise processed), and the possible displacement (for reasons of flavor alteration by cooking) of more nutritionally beneficial foods by less nutritionally beneficial foods, as possible offsets to that possible benefit require consideration.
Also, now that modern farming allows us to grow a sufficiency of foods that can be eaten very suitably from all sensible angles raw, cooking may well now, in this phase of our evolution, be a negative thing, not a positive thing (if ever it was) – overall, I am strongly inclined to think that this is the case.]

(RW, CF) "(p177) But for women, the adoption of cooking has also led to a major increase to their vulnerability to male authority."
(DV) [Probably not – the vulnerability has been there ever since women needed defense for themselves and their children, from predators and other men.]
(RW, CF) "Men were the greater beneficiaries."
(DV) [Maybe not, as all the subservience and differential drudgery was probably well in place *long* before the advent of cooking. In fact, if cooking involves less work than some other forms of food preparation, including of some other foods that might have been eaten instead in the absence of cooking, which it may well do, then women may have benefited at least as much as men.
The use of fire as a weapon also requires consideration in this regard.]
(RW, CF) "Cooking freed women's time and fed their children, but it also trapped women into a newly subservient role enforced by male-dominated culture. Cooking created and perpetuated a novel system of male cultural superiority."
(DV) [Probably very little new about it – just more of pretty much the same, with a new aspect added…]
(RW, CF) "It is not a pretty picture."
(DV) [Indeed, and, in addition to female woes, least of all the cage-farming of 'food' animals, which were it not for cooking would probably be foods less preferred than various plant food protein sources…]

Detailed review of Catching Fire
Chapter 8 The Cook's Journey (p 179-194)

(RW, CF) "(p180) Cooked food, being soft, enables mothers to wean their young early. During human evolution, early weaning would have allowed a mother to recover her body condition rapidly after birth, promoting a shorter interval between births. … …"
(DV) [Obviously enough, this is not necessarily an advantage: the carrying through pregnancy, and the care and feeding of (multiple) (very) young can be just as draining as, or more draining than, more prolonged breast-feeding – prolonged breastfeeding delays the ability for subsequent conception of the next child – understanding this crucial fundamental is necessary for almost any related discussion to be sensible…]

(RW, CF) "(p180) Mothers who weaned their babies early would have larger families than before, an infant and a toddler side by side. The advantages of grandmothers and other kin would have increased. … …"
(DV) [Yes, they'd have been more needed, as the family structure would be more vulnerable than before without them…]
(RW, CF) "By generating easily donated gifts of cooked food that are useful for the young, the new system of processing food would have favored cooperative tendencies in rearing families."
(DV) [Though this could be true, it is just as true for processed raw food, and for various types of unprocessed raw food, which were almost certainly available for sharing long before cooked food was…]

(RW, CF) "(180) Cooking would also have reduced the difficulties of finding enough to eat during the poorest seasons, when even now hunter-gatherers routinely find conditions hard."
(DV) [Probably true, but it needs to be borne in mind that, as evidenced by documentation of historical human dietary practice in times of food shortage, items that privileged modern westerners consider insufficiently tasty to be edible are in fact edible. What may have happened, in agreement with Wrangham's theorizing insofar, is that our ancestors may have eaten more during periods of

food availability and so put on more body fat to tide them over the periods of food shortage. Though of course any animal will expand its distribution to the limit of the environments ability to sustain a population, no matter how sparse, at all.]

(RW, CF) "(p180-181) The notion of cooked food making life easier challenges the thrifty-gene hypothesis, which claims that because the environments of our hunter-gatherer ancestors were highly seasonal, we are adapted to periods of feast and famine. Accordingly, ancestral humans supposedly digested and stored food in their bodies with exceptional efficiency. The thrifty-gene hypothesis suggests this efficiency was a useful adaptation when starvation was a consistent threat but is responsible for obesity and diabetes in many modern environments. The cooking hypothesis suggests a different idea: during our evolution, our use of cooked food would have left us better protected from food shortages than the great apes are, or than our noncooking ancestors were. It implies that humans easily become obese as a result of eating exceptionally high-energy, calorie-dense food, rather than from ancient adaptation to seasonality. Great apes become obese in captivity on a rich diet of cooked food."

(DV) [No, the thrifty-gene theory is not challenged at all by any extent that cooked food "made life easier" (which I think really only equates to increasing the population density or range) - our ancestors may have eaten more during periods of food availability, due to the increased palatability of foods, otherwise only marginally acceptable by taste, wrought by cooking and so put on more body fat to tide them over the periods of food shortage. Again, any animal population will expand its distribution to the limit of the environment's ability to sustain any population, no matter how sparse, and this includes population sink areas accommodating population spilling out of some more hospitable population generating area.]

(RW, CF) "(p182) When our ancestors first controlled fire, they would not have needed it for warmth, though fire would have saved them some energy in maintaining body temperature."

(DV) [The truth of this statement is very much dependant on just when our ancestors first controlled fire, because the distribution of our ancestors across the planet has differed over time, with occupation of regions having colder climates generally coming later in time. The most relevant reason this statement is generally largely true, and one that Wrangham has not recognized the great importance of to the question of the evolution of the human diet, is that our ancestors probably used animal skins/ pelts for warmth from quite early times, regardless of whether they found the associated meat (raw and often decomposing if not frankly rotting) palatable enough to eat (which would depend on how hungry they were, which would depend on location and season). Put another way, that animal skins/ pelts were probably more necessary for warmth than the meat was for nutrition, even though the two needs are more likely to be more met from the same animal source the more marginal and naturally inhospitable (coldness of climate is one major factor in this) the area occupied was – The Inuit being one of the best examples of this, at one of the very extremes of marginal and inhospitable, regards plant foods, areas ever occupied by humans or our ancestors.]

(RW, CF) "(p183) But *Homo erectus* could have lost their hair only if they had an alternative system for maintaining body heat at night. Fire offers that system."
(DV) [See my comment immediately above – and it is fatal to Wrangham's theorizing here that he is either unaware of such a basic and obvious alternative to his proposition, or has elected to not so much as mention it, much less properly address it here…]
(RW, CF) "Once our ancestors controlled fire, they could keep warm even when they were inactive. The benefit would have been high: by losing their hair, humans would have been better able to travel long distances during hot periods, when most animals are inactive. They could then run for long distances in pursuit of prey or to reach carcasses quickly. By allowing body heat to be lost, the control of fire allowed extended periods of running to evolve, and made humans better able to hunt or steal meat from other predators."
(DV) [The point is more likely to be that they would have been able to walk long distances without losing and needing as much water, in search of foods that were more likely to be most often of

plant origin. And to run from predators, given enough of a head start. And yet again, it is very fatal to Wrangham's theorizing here that he is either unaware of such basic and obvious alternatives to his proposition, or has elected to not so much as mention it, much less properly address it here…]

(RW, CF) "(p184) Even our ancestors' emotions are likely to have been influenced by a cooked diet. Clustering around a fire to eat and sleep would have required our ancestors to stay close to one another. To avoid lost tempers flaring into disruptive fights, the proximity would have demanded considerable tolerance."
(DV) [Wrangham fails here to properly differentiate clustering around a fire to eat cooked food, from clustering around it to eat food *otherwise* prepared, from clustering around it to sleep. He has also failed to mention a crucially important point in that while a campfire serves to keep dangerous non-human predators at bay, it serves only to signify position and attract any dangerous human adversaries who may be near enough by, and who, as users of fire themselves, rather than being scared away by it would only take advantage of its usefulness, not only in preliminary location of the potential target, but also insofar as the attacker from whatever distance is better able (on average) to see the victim near the fire than is the victim able to see the attacker. For this reason, less watching by sentries is necessary regarding non-human/hominin predators, but more watching by sentries may be necessary regarding other human/hominins – and even though attack by other human/hominins aided by one's campfire might be less frequent than attack by non-human/hominins scared by one's campfire, the price for long-term generational survival through natural selection may well have been constant vigilance (as well as some degree of quietness) … and the tolerance and correlate cooperation needed to maintain that vigilance (and quietness, and function as a tightly-bonded group in any combat necessary against other human/hominin groups…]

(RW, CF) "(p184-185) The first cooks probably experienced a similar process. Among the eaters of cooked food who were attracted to a fireside meal, the calmer individuals would have more comfortably accepted others' presence and would have been less likely to irritate their companions. They would have been chased away less often, would have had more access to cooked food, and would have passed on more genes to succeeding generations than the wild-eyed and intemperate bullies who disturbed the peace to the point that they were ostracized by a coalition of the calm. A version of this system had probably already started before cooking, when groups of habilines clustered about a meat carcass."

(DV) [Indeed, a version of this system had probably already started *long* before cooking, way back to 6,000,000 years ago when our line first diverged from the chimpanzee's line and spent much more time on the ground in more open areas, in the larger and well-cooperating groups that would have been necessary for survival, having very, very little to do with groups of habilines clustered about a meat carcass, on those less frequent and probably rather uncommon occasions when they were so very hungry as to be disposed to do so…]

(RW, CF) "(p185) If the intense attractions of a cooking fire selected for individuals who were more tolerant of one another, an accompanying result should have been a rise in their ability to stay calm as they looked at one another, so they could better assess, understand, and trust one another. Thus the temperamental journey toward relaxed face-to-face communication should have taken an important step forward with *Homo erectus*."

(DV) [Chimpanzees are not noted to rely *heavily* on cooperation for defense against either predators or other groups of chimpanzees, notwithstanding that they do form alliances with others within their own groups.

It seems obvious enough to me that on the other hand, even (and probably enough particularly) way back to 6,000,000 years ago when our line first diverged from the chimpanzee's line and spent much more time on the ground in more open areas, larger and much more well-cooperating groups than those of chimpanzees would very probably have been necessary for survival, regarding protection from a multitude of very formidable terrestrial predators. Cooperation, very importantly probably including that necessary for maintaining as high a level of quietness as possible for a large fraction of each day, seems to me to be an obvious requirement for the survival of a smaller, slower, weaker, slowly-reproducing, non-naturally-camouflaged, only primitively though somewhat effectively weaponized, and otherwise generally very vulnerable animal…

As *Homo erectus* was a much bigger animal than its earlier ancestors, it could probably afford, insofar (without any discussion of hominid population increases due to more effective defense from predators, leading to increased pressure on food resources, which is not a trivial issue, and, crucially it seems to me, must have become very important at some time(s) in our evolutionary history (including but not limited to the advent of agriculture), which Wrangham does not discuss at all here in his book) a little *less* cooperation than its smaller, slower (on the ground if not in the trees) and weaker hominid ancestors.

In this regard the evolution of the human supra-laryngeal vocal tract, which only very recently (no earlier than 400,000 years ago (Capasso, Michetti, D'Anastasio (2008) and possibly as late as 40,000 years ago (Donald 1991)) was such that the full modern human range of speech sounds could be produced (bear in mind that modern great apes are so limited in the range of vocal sounds that they can produce, that it is not physiologically possible for them to speak even closely to the way we can, and therefore impossible for them to be taught to communicate with us by speech, and sign language must be used instead…), and that this is very much related to the shape of the human mouth cavity and jaws, which Wrangham does not so much as mention, much less address properly in Catching Fire….

There are at least several very important possible factors other than cooking, that are very plausible alternatives to cooking's vaunted role in the evolution of human behaviour, and physiology. That Wrangham has not properly elaborated these very plausible alternative explanatory factors here in his book is very poor science…]

(RW, CF) "(p187) Climate change appears responsible for the extinction of australopithecine species. Africa began getting drier about 3,000,000 years ago, making the woodlands a harsher and less productive place to live. Desertification would have reduced the wetlands where australopithecines would have found underwater roots, such as cattails and water lilies, and they would have found fewer fruits and seeds. The species of *Australopithecus* had to adapt their diet or go extinct. Two lines survived.

One adapted by intensifying its reliance on the underground foods that had provided the backup diet of less preferred foods for australopithecines in times of food scarcity. Their descendants rapidly developed enormous jaws and chewing teeth, and are recognized in the naming of a new genus, *Paranthropus*, or the "robust" australopithecines. *Paranthropus* emerged around 3,000,000 ago, possibly descendants of *Australopithecus afarensis* or *A. africanus*. They flourished in some of the same dry woodlands as our human ancestors until 1,000,000 years ago and still looked like upright-walking chimpanzees. But even more than their *Australopithecus* ancestors, *Paranthropus* relied mainly on a diet of roots and other plant storage organs."

(DV) [This is not at all the consensus position of paleoanthropologists, for the very good reason that the archaeological evidence does not allow it to be – not least of which is that the massive jaws and teeth (the width of them being one important aspect) were at least as likely to have been used in grinding seeds, and maybe cracking open nuts, as in chewing underground storage organs (Walker 2007, Henry and Wood 2007, Lucas 2007, Ungar 2007, Teaford 2007 (1 & 2), Daegling and Grine 2007,). What abrasion marks there are on the few teeth available do not allow such an opinion as Wrangham has here presented as a fact of science, to be presented as a fact of science (Teaford 2007 (2))…

And there is the matter of possible differences in the use of naturally-formed stone and wood tools in food preparation, between the robust and gracile australopithecines, such that they may well have had different ways of dealing with some of the same foods, which Wrangham fails to so much as mention much less discuss properly…

And again, that Wrangham has not properly elaborated the very plausible alternative explanations (well enough laid out by the contributory authors in Ungar's (2007) book Evolution of the Human Diet The Known, the Unknown and the Unknowable, of which Wrangham was one of the contributing Authors), here in his book is very poor science…
Also to be considered, which Wrangham has not, is that it is possible for the fallback foods of a species to dictate the physiological peculiarities of a species more than its preferred foods, the latter which it may have in common with other species having quite different physiological peculiarities (Ungar 2007, Lambert 2007).]

(RW, CF) "(p187) The other line of descendants led to humans, and it began with meat eating."
(DV) [This is the rankest of conjecture! This is not at all the consensus, nor even the majority position of paleoanthropologists, for the very good reason that the archaeological evidence does not allow it to be. What evidence there is does not at all allow such an opinion as Wrangham has here presented as a fact of science, to be presented as a fact of science, rather, the evidence is to the contrary, where it is not simply lacking altogether…
And again, that Wrangham has not properly elaborated the very plausible alternatives (well enough laid out by the contributory authors in Ungar's (2007) book Evolution of the Human Diet The Known, the Unknown and the Unknowable), here in his book is very poor science…
And again, moreover, the modern human nutritional biochemistry, as I have elaborated at the end (p 178-183) of my reviewing of Wrangham's Catching Fire Chapter 4 above, virtually conclusively refutes this claim, Wrangham having proceeding in his theorizing without so much as a mention of this crucial aspect of the evidence relating to the central theories in Catching Fire…]
(RW, CF) "(p187-188) Australopithecines must always have been interested in eating meat when they found fresh kills, just as chimpanzees and almost every other primate are today."
(DV) [Among the apes, the most relevant primates to the human diet question(s), gorillas, orangutans and bonobos eat extremely

little or no meat.
Chimpanzees are not always interested in eating meat when they find fresh kills, as the experiments (Kortlandt 1984 citing Kortlandt 1967, Teleki 1981 citing van Lawick-Goodall 1968) placing live prey and fresh kills alongside the trails of wild chimpanzees, alongside quite a sufficient number of other reported observations of chimpanzees clearly show. Far from being always interested in opportunities to eat meat, they are mostly interested in their preferred foods, which are of plant origin, meat being mostly, even in the small proportion (5 % of calories, Stanford 1998, despite his subsequent exaggeration and sensationalization of chimpanzee carnivory in his book of 1999) of diet that it is on average, a fallback food that is eaten mostly in times or seasons of scarcity of the preferred (plant) foods (Kortlandt 1984, Yamagiwa et al 1996), when that scarcity is so great that growing juvenile chimpanzees cease to grow, a degree of shortage that in human terms is usually classed as some degree of starvation – that is, chimpanzees usually do not eat meat unless they are very, very hungry, in some degree of starvation (Hladik 1981)…
And very likely the australopithecines likewise…]

(RW, CF) "(p188) Given these challenges [(DV) physiological unsuitedness (relative physical weakness and lack of fangs and claws) for both hunting and 'power scavenging' (confronting feeding predators and driving them off the carcass to take it over)], it is unclear how australopithecines obtained access to the meat of antelope and other game animals. … …
A bold group of australopithecines might have confronted the predators with simple spears modified from digging sticks that they had used to obtain roots. … … Or maybe they threw rocks at their opponents, much as chimpanzees now sometimes scare pigs or humans with well-aimed missiles in Gombe, Tanzania."
(DV) [It seems very probable to me that, at some times of great hunger, australopithecines, in groups of sufficiently large numbers, cooperated in a similar way that they must have used in *defense* against predators, using clubs, simple spears and thrown stones, to drive predators off their antelope and other game animal carcasses and take them over ('power scavenging'), because

australopithecines would have lacked the size and speed to hunt like more naturally equipped (fangs and claws, coupled to speed and power) predators, even though their spears and clubs could have substituted somewhat for the fangs and claws.
It is possible, and even probable, that stealth and ambush could have been (and were) also used – Strum (1981), in that classic of primatology, details well enough how the Pumphouse Gang baboons in Kenya very quickly learned to cooperate in using their speed to surround (by fast running) and cut off the escape of Thomson gazelle prey, and it seems very reasonable to extrapolate from this that australopithecines, using stealth as one mainstay of survival strategy in defense against predators when not grouped together well enough for strong combative defense, would have learned to cooperate in using their stealth to surround and cut off/ambush driven prey in a manner quite analogous to that used by the Pumphouse Gang baboons as a new development to their range farm (large nearly wilderness area with cattle run on it by humans) environment having less large predators (greatly reduced in number by the farmers to protect the cattle) and accordingly more prey animals available and less danger in modifying their behaviour to include more directed and far-ranging hunting of that prey. The main difference would be the use of stealth instead of speed, the cooperation and tactics being fairly similar.]

(RW, CF) "(p189) Whatever the technique, by at least 2,600,000 million years ago, some groups were definitely getting meat from carcasses that previously only big carnivores would have eaten. Over the next few 100,000 years, impact notches and cut marks on animal bones caused by stone tools attest to habilines spending long enough in the danger zones to be able to slice the meat off the dead animals, from turtles to elephants. The result was a new and immensely beneficial food source." … …
(DV) [Wrangham has failed to understand here that the sensible quantitative analysis of the data available from around 2,600,000 years ago through to especially at least another million years after that, and on for another million years or so again, do not at all enable one to conclude that meat was anything other than a very small proportion of hominid diet, in some places, at some times.

For a start Wrangham does not indicate anywhere in his book here that he understands the nature of the discovery biases of hominin sites, such as arise from the fact that drier areas preserve hominin and associated remains whereas wetter areas do not, and the fact that bones (and particularly larger ones) as remnants of food are preserved much better than remnants of plant foods.

As Mann (1981) notes: "The archaeological record is a highly selected information source; some materials preserve very well and are routinely found during excavation, while others either are not preserved or leave such subtle traces that they are missed.

Many dietary reconstructions have been based exclusively on the recovered evidence, ignoring items that might originally have been present but that have disintegrated long since. Because vertebrate bones preserve better than plant materials or invertebrate exoskeletons, the basic diet during much of human evolution has been viewed in terms of an emphasis on a single food resource: mammals that are hunted."

And, in the entry for Paleodietary Analysis in Delson et al (Eds) (2000): "Moreover, it is notoriously difficult to calculate the proportional importance of different classes of foods consumed from archeological residues. For example, the archeological record tends to skew the evidence of prehistoric diets toward meat eating because the bones of edible animals are more likely to survive than the remains of plant foods."]

(DV) [Furthermore, Wrangham does not indicate or evidence anywhere in CF that he understands the sensible analysis of the quantitative data that there is with a view to gaining indications of the proportion of hominid diet that meat eating was. As an example of the way such data is usually grossly misconstrued I present the following extract from my book The Natural Human Diet? The Optimal Human Diet? The known and the unknown, the possible and the probable – an analysis of the evidence (p89-90), and note that it refers to a period of time, that is an archeologically very recent period of time, when hominins (humans, here) very probably did eat significantly more meat (though still only a small proportion of the diet, in some places, at some times, as I think I fairly conclusively demonstrate…) than their hominid ancestors did

millions of years ago:
{{[However, the following is an example very typical of the almost universal failure to do what tentative quantifying is readily able to be done (the maths is very simple):
Although Olsen (1989) does a good job of analyzing the probabilities that various hunting methods were used in the generating of the great pile of horse remains in the Upper Paleolithic (from at least 30,000 years before the present up to 12,000 years before the present) big game kill site in Solutre in what is now France, she unfortunately did not carry out the mathematically childish calculation of horse/human/year – most readers are unfortunately (both on her, and on their, accounts….) likely to form the notion that the accumulation of the remains of "32,000 to 100,000" horse carcasses at the site is indicative of a most ferociously carnivorous existence of the humans involved.
The calculation proceeds as follows:
Taking the middle value of 60,000 horse carcasses, over the least number of years (20,000 years), gives 3 horses killed per year.
Now, unlike the much less numerous reindeer remains at the site where only a cursory examination of 100 bones "yielded numerous examples of reliable butcher marks", only 17 of the 4,483 horse bones examined bore signs of butchery – this is 0.38%.
However, to allow for the possibility that the bones with butcher marks were much larger (10 x) than the bones that didn't have butcher marks, 0.38% is changed to 3.8%, which might well lead to a large exaggeration of the consumption of horse per person per year, but is unlikely to underestimate it.
So, 3.8% of 3 horses killed per year is 0.11 horse butchered/eaten per year.
And, this one tenth of a horse was eaten, in a year, by the group of people involved, say a group of 10 people, which is probably a minimum, the number quite possibly being twice or more that.
This gives one hundredth of a horse (the meat, not all of the carcass) eaten per person per year.
For an average horse of 400kg, very roughly 200kg of meat, one hundredth of that is 2kg of meat.
That is, 2kg of horse meat per year per person.
Or, 5 or 6 grams (.005 or .006kg) of horse meat per day per person.

And maybe it wasn't all eaten.
Maybe the hides were what they really wanted, to make warm clothes or even dwellings or tools/ accoutrements (being French as they were…) from.

And the real figure could be *much* less than this estimate which has been deliberately made conservative towards agreeing with the orthodoxy's claims of humans having a ferociously carnivorous past and away from a more predominantly vegetarian past.
Which hardly constitutes a most ferociously carnivorous ex(subs)istence.
So, it seems likely, every now and then (maybe every year or few years), probably as much for the excitement, display of martial power, and ego-trip of it as for anything else other than to get hides for clothing, and/or for some ceremonial/religious reason(ing) (the horses were taken in summer, when other food was plentiful, the reindeers were taken in winter, when other food was not plentiful) they drove some horses up the canyon to the dead end, killed them (maybe released some), ate a little, and left the rest to rot.
Violent, yes, carnivorous, much less so, it seems, but without this little bit of basic maths to calculate the likely actual contribution to *daily* diet of this activity carried out over tens of thousands of years the image very falsely given is one of predominant carnivory….]}}]

(DV) [And, moreover, modern human nutritional biochemistry, as I have comprehensively enough elaborated in the last few pages (p 178-183) of my review of Wrangham's Catching Fire's Chapter 4 above, virtually conclusively indicates that meat has been no more than a very, very minor proportion of our ancestral diets as resulted in the evolution of our present biochemistry.]

(RW, CF) "(p189) Habilines' chewing teeth were similar in size and shape to those of australopithecines, showing a continuing commitment to the same plant foods, including raw roots and corms during the most difficult seasons, and such items as soft seeds and fruits when they could find them."
(DV) [Indeed, and note that despite, as Wrangham notes in his

book here, that evolutionary change can be accomplished quite quickly (for example the adult retention of lactase for the digestion of milk sugar appearing only in the last several thousand years in some human populations having significant milk products in the post-infancy diet), there was no trend to humans evolving back the elongated canine teeth that our ancestors lost after our divergence from the evolutionary line of the chimpanzees. In combat, the teeth could be replaced by weapons, but in feeding, elongated teeth are used in carnivory by all animals, in particular by the chimpanzees and baboons, notwithstanding that bonobos, gorillas, and orangutans do not use their elongated canine teeth for carnivory to any significant extent.]

(RW, CF) "(p189-190) Probably habilines prepared nuts by smashing them to expose the edible seeds, as chimpanzees do. It is doubtful that habilines could process plant foods by any techniques that were much more elaborate than pounding."
(DV) [Grinding, as a potentially very important only slight modification (in its most primitive form) of cracking and pounding, requires careful consideration here, which Wrangham seems unaware of. Likewise, threshing, collecting, carrying, storage, drying, and soaking. Likewise peeling and pulping, and juice extraction even if only by spitting out fiber as some gatherer-hunters/ hunter-gatherers have been reported to do.]
(RW, CF) "(p190) Almost all the methods hunter-gatherers use to improve the nutritional value of plant foods involve fire, because heat is needed to gelatinize starch."
(DV) [No, threshing, collecting, carrying, storage, drying, and soaking also improve the nutritional value of plant foods, without the use of fire. Likewise peeling and pulping, and juice extraction including that only by the spitting out of fiber.
Even for those methods that hunter-gatherers use to improve the nutritional value of plant foods that do involve fire, it is certainly not only because heat is needed to gelatinize starch, but also because fire use carries technological and pseudo-magical status, and because of the flavor improvement of some things, and because of warmth including that from food eaten.]
(RW, CF) "(p190) Until fire was controlled, habilines would have

been stuck with eating raw plant foods whose caloric value could not be much improved by cold processing."
(DV) [Incorrect – cracking, threshing, crushing, grinding, pulping, mashing, peeling, and juice extraction including that only by the spitting out of fiber, drying, and soaking also significantly improve the nutritional value of plant foods, without the use of fire..]

(RW, CF) "(p190) The big question for the habilines is not how they tended fire, but how they would regularly have obtained it. …
…
(p191-192) Anthropologists caution that the sparks produced by many kinds of rock are too cool or short-lived to catch fire. But when pyrites, a common ore containing iron and sulfur, are hit against flint, the result is a set of such excellent sparks that pyrites and flint are standard components of fire-making kits used by hunter-gatherers from the arctic to Tierra del Fuego. If a particular group of habilines lived in an area exceptionally rich in pyrites, they could have found themselves inadvertently making fire rather often.
The steps to managing fire need not have involved the difficult process of deliberately making it. Here is an alternative scenario: during the tens of thousands of generations between the origins of habilines (at least 2,300,000 years ago) and *Homo erectus* (at least 1,800,000 years ago), from time to time the sparks resulting from habilines pounding rocks could have accidentally produced small fires in adjacent brush. Perhaps cocky juvenile habilines dared to grab the cool end of a branch and tease one another with the smoldering twigs or blazing leaves, much as young chimpanzees playfully bully one another with sticks they use as clubs. Adults learned the effect on one another of waving a burning log. The process was then transferred to the serious job of frightening lions, sabertooths and hyenas, similar to how chimpanzees use clubs against leopards. At first, the fires went out. But over time, when sparks happened to start a fire, habilines learned that it was worth their while to keep it going. They cultivated fire as a way to help them defend against dangerous animals."
(DV) [And for technological and pseudo-magical status…. Some relevant detail here might be the different utility of fire to larger

groups of armed habilines (where it might not have been much of an advantage), versus smaller groups including children (where it might have been a significant advantage), grouping maybe being dependent on what foraging strategies best suited the particular environment, modified again by whether or not relationships with neighboring groups of habilines were friendly or not (regarding vulnerability to being located in order to be attacked during either night or day)….]

(RW, CF) "(p192) There are other possibilities. The climate was becoming increasingly dry. Natural fires could have become more frequent. Perhaps people walked behind brush fires looking for cooked seeds. Maybe they obtained fire from seeds that burned slowly after being struck by lightning; a eucalyptus tree can smolder for eight months. Perhaps there was a permanent natural source somewhere in Africa, like the gas-fired strip of flame that has been burning nonstop near Antalya in southwestern Turkey ever since Homer recorded it in the Iliad almost three thousand years ago. … …"
(DV) [Volcanoes also warrant a mention here.]

(RW, CF) "(p192-193) But if there were a natural source of fire, such as sparks, there would have been no need to learn to make fire, because it could be taken from nature again and again, and eventually from other groups."
(DV) [This is illogical – fire coming from sparks (from struck stones) strongly implies fire very soon coming to be made deliberately from sparks….]
(RW, CF) "the chance of a rainstorm extinguishing every fire in the neighborhood would have become vanishingly small."
(DV) [Maybe, if some fires were successfully protected from extinguishment by any widespread rain such as does occur seasonally in many places…]
(RW, CF) "Among Australian aborigines, groups that lost their fire from a drenching rain or flood would refresh their supply from neighbors, who would expect something in return, such as quartz flakes or red ochre."
(DV) [Protection of fire from extinguishment would be necessary

in the monsoonal north of Australia, such as Arnhem Land, if long periods without fire were to be avoided.]

(RW, CF) "(p193) Keeping a fire lit would have been a big achievement, but logs are easy to keep aflame when people are moving. Hunter-gatherers regularly carry fire in the form of a burning log. As long as the carrier is walking, the fire is well oxygenated and the log continues to smolder. When people stop, they start a small fire within a few minutes by adding a few sticks to the smoldering log and blowing."

(RW, CF) "(p193-194) Once they kept a fire alive at night, a group of habilines in a particular place occasionally dropped morsels by accident, ate them after they had been heated, and learned that they tasted better."
(DV) [Firstly, the first tasting of cooked food almost certainly occurred following forest fires, not at campfires.
Secondly, it is very likely, consistent with trials on captive apes, that not all cooked food tasted better than the raw food – in the case of meat the cooked food probably tasted better; in the case of many fruits and some vegetables, it probably didn't…]

(RW, CF) "(p194) Repeating their habit, this group would have swiftly evolved into the first *Homo erectus*. The newly delicious cooked diet led to their evolving smaller guts, bigger brains, bigger bodies, and reduced body hair; more running; more hunting; longer lives; calmer temperaments; and a new emphasis on bonding between females and males."
(DV) [It is poor science by Wrangham here not to have properly presented the quite obvious and very (more, in the opinion of many) plausible alternative explanation that an increase in food macronutrient (carbohydrate, fat, and protein, specifically such as to combine in providing calories (energy)) density such as could be provided by some combination of firstly finding, accessing (e.g. using digging sticks to access starchy roots) and gathering more macronutrient dense foods; secondly by processing them in various ways with naturally formed and primitively manufactured stone and wood tools, for example: cracking open oily nuts and seeds,

grinding cereal grains and other seeds, peeling and pulping starchy roots, pulping and juice extracting starchy roots and fruits, pulping fruits, drying foods with or without preliminary pulping or crushing, soaking dried or crushed food; and thirdly by transport and storage of processed and unprocessed foods.

When an animal (habilines, here) is fangless, clawless, non-camouflaged, relatively small, relatively weak, relatively slow, and not well-protected by its niche (more open countryside, with less use being made of trees for protection) in its physical environment, an increase in size is fairly obviously going to be an advantage for survival, and to be selected for in evolution. Likewise, increased height for better looking over grasses (if average or significant grass height is such as to be a driving factor) to maintain vigilance and locate food. The very largest animals do not eat a calorically concentrated diet.]

(RW, CF) "(p194) The softness of their cooked plant foods selected for smaller teeth, "

(DV) [It is very poor science here that Wrangham has not addressed also the possibility that processing of food by methods other than fire, namely cracking, crushing, grinding, peeling, cutting, mashing/pulping, drying and soaking, and juice extraction, particularly as all of these methods probably were in common routine use long before fire was, contributed to selection for smaller teeth.

Also, the fact that modern humans, like our pre-human ancestors, have thicker and stronger tooth enamel than any of the apes should be addressed.

Likewise, it is very poor science here that Wrangham has not addressed also the possibility that the evolution of the human supra-laryngeal vocal tract from something like that of the apes (who are unable to make any more than a small fraction of the vocal sounds that modern humans are able to, and who are therefore unable to communicate anything other than the most basic things through vocalization, such that they have to be taught to communicate with humans by sign language instead of speech), to that of the modern humans such as to give us the ability that we have for vocal communication of highly complex and accurate

information and ideas, which is obviously an evolutionary advantage when coupled to intelligence, contributed to selection for smaller teeth.
Moreover, no evidence has been presented that the use of cooking was any more than a very small fraction, if any, of the evolutionary drives to the evolution of the modern human teeth.]

(RW, CF) "(p194) the protection fire provided at night enabled them to sleep on the ground and lose their climbing ability, "
(DV) [It is poor science here by Wrangham not to have properly presented the quite obvious and very (more, in the opinion of many) plausible alternative explanation that, evidenced by the evolution of humanoid (hominid) feet very early in evolution from the common ancestor of humans and chimpanzees, even very early hominids were spending a lot of time on the ground (and relatively treeless ground), which they could not do as the fangless, clawless, non-camouflaged, relatively small, relatively weak, relatively slow animals they were, without a high level of cooperation by members of a group (made up of more than just one nuclear family, probably of several), even with the basic weapons which they almost certainly had and used from very early in evolution, early hominids slept on the ground very often, using staggered sentry watches and the same sort of cohesive group activity to defend themselves when awakened to do so by the sentries, if not then climbing the trees for escape or temporary refuge.
Also, it is poor science here by Wrangham to have not understood or failed to mention that whereas campfires are a defense against animal predators, they are a liability against enemy hominids/ humans, because they identify the location of the campers, and also provide a visibility advantage to the incoming enemies in formulating attacks.]

(RW, CF) "(p194) and females likely began cooking for males, whose time was increasingly free to search for more meat and honey."
(DV) [Increasingly free to escape routine domestic drudgery and engage in more exciting activities, even though some of this involved the swapping of domestic drudgery for the risk of death or

injury in maintaining (defending) or acquiring territories, notwithstanding that this risk could be reduced by infanticide limiting population numbers such as to reduce the drive of resource (food) scarcity towards territorial conflict.]

Detailed review of Catching Fire
Epilogue The Well-Informed Cook (p 195-207)

(RW, CF) "(p200) For example, the energy value of different types of protein is known to vary: egg protein produces 4.36 kcal/gram, whereas brown rice protein produces 3.41 kcal/gram, and so on."

(RW, CF) "(p201) New general factors have also been proposed for dietary fiber (or nonstarch polysaccharides), which are so much less well digested than other carbohydrates that they deserve a lower energy value than 4 kcal/gram; a figure of 2 kcal/gram has been proposed."

(RW, CF) "(p202) The first problem is that the Atwater convention does not recognize that digestion is a costly process. When we eat, our metabolic rate rises, the maximum increase averaging 25 percent. The corresponding figures for fish (136 %) and for snakes (687%) are vastly higher, showing that humans pay less for digestion than other species, presumably due partly to our food being cooked."
(DV) [Mostly what these differences indicate is not differences in the cost of digestion, but differences in basal metabolic rate – related to 'warm-bloodedness', 'cold-bloodedness' and intermediateness of animal species….]

(RW, CF) "(p202) Protein costs more to digest than carbohydrates, while fat has the lowest digestive cost of all macronutrients.
In a 1987 study [cites Sims and Danforth], people eating a high-fat diet achieved the same weight gain as others eating almost 5 times the number of calories in the form of carbohydrate."
(DV) [The figure of 5 seems a gross exaggeration of what difference there actually is. My personal experience of body fat gain from eating, for example, bananas (or other of the more highly calorific fruits) in excess of my energy requirements is not in accord with Wrangham's claim here. Nor is the Atkins Diet experience (though I am far from being an advocate of the Atkin's diet), notwithstanding that people on the Atkins Diet probably do eat less calories due to their hunger being less and more stable than

when on a nearly fat-free diet or fat-free meal including diet.]

(RW, CF) "(p203) Based on animal studies, we can expect that the costs of digestion are higher for tougher or harder foods than softer foods; for foods with larger rather than smaller particles; for food eaten in single large meals rather than in several small meals; and for food eaten cold rather than hot."
(DV) [Absolute digestibility is just as likely to be the issue here as the cost of digestion.]

(RW, CF) "(p203) Individuals vary too. Lean people tend to have higher costs of digestion than obese people. Whether obesity leads to a low cost of digestion or results from it is unknown."
(DV) [It is fairly obviously very unlikely that obesity results in increases in absorption or a "low cost of digestion" – in general the result is *resistance* to the absorption of more of the excess nutrients – for example insulin resistance, for which there are probably as yet unmeasured correlates in GIT absorption – which would need to be compared within the same person in lean and obese conditions, rather than between randomly selected lean and obese individuals, for reasons that are, I would hope, obvious…]
(RW, CF) … … For the same number of measured calories, an obese person, having a lower digestive cost, will put on more pounds than a lean person."
(DV) [It is at least as likely to be differences in absorption, rather than differences in the cost of digestion per se, that is in effect – that is, that the digestive tracts may do much the same amount of work, but in the genetically leaner person less is absorbed…]

(RW, CF) "(p203-204) When A.L. Merrill and B.K. Watt introduced the Atwater specific-factor system in 1955, they noted specifically that the digestibility of a grain is affected by how finely it is milled. More extensively milled flour might be completely digested, whereas less milling could lead to 30 % of the flour being excreted unused. … …"
(DV) [It should be noted here that the meal context may be very important here – is the flour or its food product eaten by itself and thoroughly chewed and insalivated, thereby maximizing digestion,

or is it eaten covered with butter and amongst other slippery or watery substances, and so swallowed without much chewing and insalivation, so as to reduce its digestion?...]

(RW, CF) "(p204) To complicate matters further, the digestibility of the same items varies according to the food context in which it is consumed. For example, protein digestibility tends to be lower when the protein is part of a high-fiber food than when part of a low-fiber food. … …
Very few studies use the only appropriate measure, ileal digestibility, which takes the sample of unused food at the end of the small intestine, rather than at elimination from the body."
(DV) [This is incorrect, the physiologically relevant measure requires some inclusion of the digestion of non-starch carbohydrate (digestion that may itself liberate more digestible carbohydrate trapped within non-starch carbohydrate structures…) that does take place in the large intestine and produces the small-chain fatty acids that the colon and particularly the descending colon, and probably other body tissues to some significant extent, use for energy.]

(RW, CF) "(p204-205) Essentially, nutrition science is faced with choosing between the immense effort of accumulating nutritional-value data that are difficult to quantify but accurate, on the one hand, or using easily quantified but physiologically unrealistic measures, yielding only a rough approximation of food value. Given the difficulty of acquiring the actual, contextually adjusted nutritional values of individual foods (and combinations of foods), the general public is provided with estimates of food values that do not reflect the realities of the digestive process."
(DV) [But that are still a lot better than nothing at all, despite the need for the general public to know some of the basic modifying factors and contexts, such as chewing, insalivation, lubrication and coating by food oil, moistening and dilution with water, the order of eating of food groups within a meal, emotional/arousal state, and physiological hunger vs. hedonistic habit, for a start…]
(DV) [hedonistic *here* pertaining to lower-level more animalistic functions – hedonism can be very usefully considered to include the pleasure derived from understanding and improving the world.]

References for critical review of Catching Fire:

Bilsborough S, Mann N (2006) A review of issues of dietary protein intake in humans. Int J Sports Nutr Exerc Metab Apr;16(2):129-52 (abstract only read).

Blumenschine RJ, Pobiner, BL. Chapter 10 Zooarcheology and the ecology of Oldowan hominin carnivory. In, Ungar PS. (Ed) (2007) Evolution of the human diet The known, the unknown, and the unknowable. Oxford University Press.

Bouguerra F, Hajjem S, Guilloud-Bataille M, Khlat M, Khaldi F, Bennaceur B (1998) Breast feeding effect relative to age of onset of celiac disease. Arch Pediatr 5(6): 621-621.

Bunn HT. Chapter 11 Meat made us human. In, Ungar PS. (Ed) (2007) Evolution of the human diet The known, the unknown, and the unknowable. Oxford University Press.

Capasso L, Michetti E, D'Anastasio R (2008) A *Homo erectus* hyoid bone: possible implications for the origin of the human capability for speech. Coll. Antropol. Dec; 32(4): 1007-11. (Abstract only read)

Clarke R, Halsey J, Bennett D, Lewington S (2011) Homocysteine and vascular disease: review of published results of the homocysteine-lowering trials. J Inherit Metab Dis. Feb;34(1):83-91. Epub 2010 Nov 11.

Cordain L. Chapter 19 Implications of Plio-Pleistocene hominin diets for modern humans. In, Ungar PS. (Ed) (2007) Evolution of the human diet The known, the unknown, and the unknowable. Oxford University Press.

Clutton-Brock TH and PH Harvey (1980) Primates, brains and ecology. Journal of the Zoological Society of London. 190: 309-323.

Cutress TW (2003) Dental caries in South Pacific populations: a review. Pac Health Dialog Mar;10(1):62-7 (abstract only read).

Daegling DJ, Grine FE. Chapter 6 Mandibular biomechanics and the paleontological evidence for the evolution of human diet. In, Ungar PS. (Ed) (2007) Evolution of the human diet The known, the unknown, and the unknowable. Oxford University Press.

Deacon TW (1990) Fallacies of progression in theories of brain-size evolution. International Journal of Primatology. 11: 193-236.

Delson E, Tattersall I, Van Couvering JA, Brooks AS. (Eds) (2000) Encyclopedia of human evolution and prehistory (2nd Ed). Garland Publishing Inc, New York.

Donald M (1991) Origins of the Modern Mind. Three stages in the evolution of culture and cognition.Harvard University Press.

Eaton SB, Konner M, Shostak M (1988) Stone agers in the fast lane: chronic degenerative diseases in evolutionary perspective. American J Med 84: 739-749.

Eaton SB, Shostak M, Konner M (1988) The Paleolithic Prescription. A program of diet & exercise and a design for living. Harper and Row, New York.

Franck F (1959) Days with Albert Schweitzer A Lambarene landscape. Henry Holt & Co.

Fontana L, Shew JL, Holloszy JO, Villareal DT (2005) Low bone mass in subjects on a long-term raw vegetarian diet. Arch Intern Med Mar 28; 165(6):684-9.

Gaqne G (1993) Mouth diseases in a prehistoric agricultural population of northeastern North America. J Can Dent Assoc Aug;59(8):686-92 (abstract only read).

Harding RSO, Teleki G. Chapter 1 Introduction. In, Harding RSO and Teleki G (Eds) (1981) Omnivorous Primates Gathering and hunting in human evolution. Columbia University Press.

Hart D, Sussman RW (2005) Man the Hunted: Primates, Predators and Human Evolution. Westview Press, New York. (glimpsed only)

Hay G, Johnston C, Whitelaw A, Trygg K, Refsum H (2008) Folate and cobalamin status in relation to breastfeeding and weaning in healthy adults. Am J Clin Nutr 88: 105-114.

Hayden B. Chapter 10 Subsistence and ecological adaptations of modern hunter/gatherers. In, Harding RSO, Teleki G (Eds) (1981) Omnivorous primates Gathering and hunting in human evolution. Columbia University Press. New York

Henry AC, Wood B Chapter 2 Whose Diet. In, Ungar PS. (Ed) (2007) Evolution of the human diet The known, the unknown, and the unknowable. Oxford University Press.

Hladik C-M. Chapter 7 Diet and the evolution of feeding strategies among forest primates. In, Harding RSO, Teleki G (Eds) (1981) Omnivorous primates Gathering and hunting in human evolution. Columbia University Press. New York.

Isaac GLI, Crader, DC. Chapter 3 To what extent were early hominids carnivorous? An archaeological perspective. In, Harding RSO and Teleki G (Eds) (1981) Omnivorous Primates Gathering and hunting in human evolution. Columbia University Press.

Koebnick C, Strassner C, Hoffmann I, Leitzmann C (1999) Consequences of a long-term raw food diet on body weight and menstruation: results of a questionnaire survey. Ann Nutr Metab 43(2):69-79.

Kortlandt A (1984) Chapter 5 Habitat richness, foraging range and diet in chimpanzees and some other primates. In, Chivers DJ, Wood BA, Bilsborough A (1984) Food acquisition and processing in primates. Plenum Press, New York.

Lambert JE. Chapter 17 Seasonality, fallback strategies, and natural selection. In, Ungar PS. (Ed) (2007) Evolution of the human diet The known, the unknown, and the unknowable. Oxford University Press.

Lee RB, de Vore I (Eds) (1968) Man the Hunter. Wrenner-Greg Foundation for Anthropological Research. Transaction Publishers, Rutgers U, New Jersey. (glimpsed only)

Lee S., Wolpoff MH (2003). The pattern of evolution in Pleistocene human brain size. Paleobiology, 29(2), 186–196.

Lucas PW. Chapter 3 The evolution of the hominin diet from a dental functional perspective. In, Ungar PS. (Ed) (2007) Evolution of the human diet The known, the unknown, and the unknowable. Oxford University Press.

Mann AE. Chapter 2 Diet and human evolution. In, Harding RSO and Teleki G (Eds) (1981) Omnivorous Primates Gathering and hunting in human evolution. Columbia University Press.

McLaren DS (1981) The luxus vitamins – A and B12. Am J Clin Nutr 34: August p1611-1616.

Milligan LA, Bazinet RP (2008) Evolutionary modifications of human milk composition: evidence from long-chain polyunsaturated fatty acid composition of anthropoid milks. Journal of Human Evolution 55: 1086-1095.

Millward DJ (2001) Protein and amino acid requirements of adults: current controversies. Can J Appl Physiol 26 Suppl: S130-40. (Abstract only read).

Milton K "Diet and Primate Evolution." In, Goodman A, Dufour D, Pelto G (Eds) (2000) Nutritional Anthropology: Biocultural Perspectives on Food and Nutrition. (p46-54). Mayfield Publishing Company, Mountain View, CA

Mobley CC (2003) Nutrition and dental caries. Dent Clin North Am Apr;47(2):319-36 (abstract only read).
Muller MN (2007) Book Reviews (Hart D, Sussman RW (2005) Man the Hunted: Primates, Predators and Human Evolution) Am J Phys Anthrop 132:000-000(?)

Nestle M (1999) Animal v. plant foods in human diets and health: is the historical record unequivocal? Proc Nutr Soc May;58(2):211-8.

Oftedal OT (1981) Milk protein and energy intakes of suckling mammalian young. A comparative study. PhD Diss. Cornell U, Ithaca, New York.

Olsen SL (1989) Solutre: A theoretical approach to the reconstruction of Upper Paleolithic hunting strategies. Journal of Human Evolution 18, 295-327.

Panter-Brick C (2002) Sexual division of labor: energetic and evolutionary scenarios. Am J Hum Biol Oct;14(5):627-40 (abstract only read).

Perry GH, Dominy NJ, Claw KG, Lee AS, Fiegler H, Redon R, Werner J, Villanea FA, Mountain JL, Misra R, Carter NP, Lee C, Stone AC (2007) Diet and the evolution of human amylase gene copy number variation. Nat Genet Oct;39(10):1256-60.

Peters CR. Chapter 13 Theoretical and actualistic ecobotanical perspectives on early hominin diets and paleoecology. In, Ungar PS. (Ed) (2007) Evolution of the human diet The known, the unknown, and the unknowable. Oxford University Press.

Potts R. Section 9.1 The hominid way of life. In, Chapter 9 Early human behaviour and ecology. In Jones S, Martin R, Pilbeam D, Bunney S (Eds) (1992) The Cambridge Encyclopedia of Human Evolution. Cambridge University Press.

Price WA (1939) Nutrition and Physical Degeneration A Comparison of Primitive and Modern Diets and their Effects.

Reed KE, Rector AL. Chapter 14 African Pliocene paleoecology Hominin habitats, resources and diets. In, Ungar PS. (Ed) (2007) Evolution of the human diet The known, the unknown, and the unknowable. Oxford University Press.

Richards MP (2002) Review A brief review of the archeological evidence for Paleolithic and Neolithic subsistence. Eur J Clin Nutr 56:1-9.

Ripley S (1984) Chapter 2 Environmental grain, niche diversification and feeding behavior in primates. In, Chivers DJ, Wood BA, Bilsborough A (1984) Food acquisition and processing in primates. Plenum Press, New York.

Roebroeks W, Villa P (2011) On the earliest evidence for habitual use of fire in Europe. Proc Natl Acad Sci USA Mar 29:108(13):5209-14.

Segal I (2002) Physiological small bowel malabsorption of carbohydrates protects against large bowel diseases in Africans. J Gestroent Hepat 17: 249-252.

Sept J. Chapter 15 Modeling the significance of paleoenvironmental context for early hominin diets. In, Ungar PS. (Ed) (2007) Evolution of the human diet The known, the unknown, and the unknowable. Oxford.

Shea JJ. Chapter 12 Lithic archaeology, or, what stone tools can (and can't) tell us about early hominin diets. In, Ungar PS. (Ed) (2007) Evolution of the human diet The known, the unknown, and the unknowable. Oxford University Press.

Shils ME, Shike M, Ross AC, Caballero B, Cousins RJ (2006) (Eds) Modern Nutrition in Health and Disease (10th Ed), Lippincott Williams and Wilkins, New York, London.

Silberbauer G. Chapter 12 Hunter/gatherers of the Central Kalahari. In, Harding RSO, Teleki G (Eds) (1981) Omnivorous primates. Gathering and hunting in human evolution. Columbia University Press. New York

Stanford CB (1999) The hunting apes. Meat eating and the origins of human behavior. 1999 Princeton University Press.

Stanford CB (1998) Chimpanzee and red colobus The ecology of predator and prey. Harvard University Press.

Steinman B, Gitzelmann R, Van den Berghe G (2001) Chapter 70 Disorders of fructose metabolism. In, Scriver CR, Beaudet AL, Sly WS, Valle D, Childs B, Kinzler KW, Vogelstein B (Eds) The Metabolic & Molecular Bases of Inherited Disease (8th Ed). McGraw-Hill, New York.

Tack GJ, Verbeek WHM, Schreurs MWJ, Mulder CJJ (2010) The spectrum of celiac disease: epidemiology, clinical aspects and treatment. Nat Rev Gatroenterol Hepatol 7: 201-213.

Teaford MF. Chapter 5 What do we know and not know about diet and enamel structure? In, Ungar PS. (Ed) (2007) Evolution of the human diet The known, the unknown, and the unknowable. Oxford University Press.

Teaford MF (1). Chapter 4 Dental Functional Morphology. In, Ungar PS. (Ed) (2007) Evolution of the human diet The known, the unknown, and the unknowable. Oxford Univ. Press.

Teaford MF. (2) Chapter 7 What do we know and not know about dental microwear and diet? In, Ungar PS. (Ed) (2007) Evolution of the human diet The known, the unknown, and the unknowable. Oxford University Press.

Teleki G. Chapter 9 The omnivorous diet and eclectic feeding habits of chimpanzees in Gombe National Park, Tanzania. In, Harding RSO, Teleki G (Eds) (1981) Omnivorous primates Gathering and hunting in human evolution. Columbia University Press. New York

Ungar PS. Chapter 4 Dental functional morphology The known, the unknown, and the unknown. In, Ungar PS. (Ed) (2007) Evolution of the human diet The known, the unknown, and the unknowable. Oxford University Press.

Ungar PS. Chapter 21 Limits to knowledge on the evolution of hominin diet. In, Ungar PS. (Ed) (2007) Evolution of the human diet The known, the unknown, and the unknowable. Oxford University Press.

Vance DJ (2010/11) The Natural Human Diet? The Optimal Human Diet? The known and the unknown, the possible and the probable - an analysis of the evidence. Multifactor Health and Education Initiative (Multifactor Health website: http://sites.google.com/site/multifactorhealth)

Veldhorst MA, Westerterp-Plantanga MS, Westerterp KP (2009) Gluconeogenesis and energy expenditure after a high-protein, carbohydrate-free diet. Am J Clin Nutr Sep;90(3):519-26.

Walker A. Chapter 1 Early hominin diets. In, Ungar PS. (Ed) (2007) Evolution of the human diet The known, the unknown, and the unknowable. Oxford University Press.

Watanabe, F (2007) Vitamin B12 sources and bioavailability. Exp Biol Med 232: 1266-1274.

Wrangham R. Chapter 16 The cooking enigma. In, Ungar PS. (Ed) (2007) Evolution of the human diet The known, the unknown, and the unknowable. Oxford University Press.

Wrangham R (2009) Catching Fire How cooking made us human. Basic Books, New York.

Yamagiwa J, Maruhashi T, Yumoto T, Mwanza N. Chapter 7 Dietary and ranging overlap in sympatric gorillas and chimpanzees in Kahuzi-Biega National Park, Zaire. In, McGrew WC, Marchant LF, Nishida T (1996) (Eds) Great Ape Societies. Cambridge University Press.

Vance ED (2011) The Normal Human Diet? The Optimal Human Diet? The known and the unknown, the possible and the probable. An analysis of the evidence. Multifactor Health and Education Initiative, Multifactor Health website http://sites.google.com/site/multifactorhealth

Appendix 1 Previous Dialogue with Prof. Wrangham on my review of his chapter The Cooking Enigma in Ungar (2007) and cursory review of the first four chapters of Catching Fire (Vance 2011)

Critical review of Prof. Wrangham's Chapter 16 The Cooking Enigma in Prof. PS Ungar's (2007) "Evolution of the Human Diet The Known, the Unknown and the Unknowable," which was done together with a brief preliminary critical review of the first four chapters of Wrangham's book "Catching Fire How Cooking Made us Human," followed by Wrangham's first response to my criticisms, my response to that, and his response to mine, are all parts of my book "The Natural Human Diet? The Optimal Human Diet? The known and the unknown, the possible and the probable – an analysis of the evidence," which is freely available from the Multifactor Health website at: http://sites.google.com/site/multifactorhealth (but surely better purchased in print…).

The reproduction here {{ between double curly brackets below }} (with identifying initials added) commences with section 1.14 of Chapter 1 of my abovementioned book, which is followed by the ensuing email dialogue between Prof. Wrangham and myself.

In which, Wrangham's words are identified by his and Prof. Peter Ungar's initials and then enclosed in quotation marks:

(RW in PU) "abc … … xyz";

and my words of commentary/ criticism are identified by my initials and then enclosed in square brackets (DV) [abc … … xyz]

{{ 1.14 Cooking

Wrangham R. Chapter 16 The cooking enigma. In, Ungar PS. (Ed) (2007) Evolution of the human diet The known, the unknown, and the unknowable. Oxford University Press.

(RW in PU) "There are two contrasting views.

The first, which is conventional wisdom, sees cooking as merely one of many extra-oral food-processing techniques (such as pounding, grinding, or drying) that can raise food quality. …… For example, it would not be considered to have led to fundamental changes in the human digestive system. …… The essential implication is that human biological evolution was not influenced in any major ways by the adoption of cooking and that evolutionists can therefore ignore it.

The radical alternative is that cooking is a core human adaptation that has importantly directed our evolution, or as Coon (1954) wrote, that cooking was "the decisive factor in leading man from a primarily animal existence into one that was more fully human." ……
Thus, the cost of digestion is a high proportion of total energy expenditure in all animals. In humans it has been measured at around 5%-15% of energy expenditure (Westerterp, 2004). ……
For instance, high-protein diets increase the cost of digestion by about 30% compared to high-fat diets (Westerterp-Plantenga et al., 1999). ……

As expected, the glycemic index is indeed consistently increased by cooking (Brand et al., 1985; Bjorck, Liljeberg and Ostman, 2000). ……

In habitats with little or no fruit, gorillas can survive by eating fibrous foods for 100% of their feeding time (Doran et al., 2002).
Chimpanzees never do so (Doran et al., 2002). ……
Thus, gorilla molars have long shearing edges compared with those of chimpanzees, and gut passage rates in gorillas are longer than in chimpanzees, allowing more opportunity for fermentation of plant fiber (Milton, 1999; Remis et al., 2001; Remis and Dierenfeld, 2004)."
(DV) [Wrangham should also have mentioned that chimps like fat in food, whereas gorillas dislike it] ……

(RW in PU) "For example, it [cooking] should increase the range of edible foods and therefore allow extension into new biogeographical zones.
Other things being equal, it should also provide a more predictable

food supply during periods of scarcity because it enables a range of otherwise inedible items to be used.
It should have further effects by softening food.
For example, it should lead to a greater ability of adults to provision infants, whose dentition is too immature to allow hard chewing, other than by giving milk.
It should likewise cause a substantial drop in the amount of time that individuals spend chewing, with large consequences for the species activity budgets (Wrangham and Conklin-Brittain, 2003)."
(DV) [But cooking also requires time; and the smoke may attract unwanted attention that then has to be guarded against; and the food being cooked is less immediately portable than it was before should exiting the area become necessary or strategic; and food can be processed for young children using naturally shaped stone and wood tools…]
(RW in PU) "Unlike the ordinary feeding pattern of any nonhuman ape, therefore, this means that a cooking population is exposed to intragroup [and intergroup…] competition over a valuable accumulated food pile.
For example, among chimpanzees by far the most valuable type of concentrated food supply is meat."
(DV) [Firstly it needs to be noted that the most valuable, and valued (sought after) foods for chimpanzees are not "concentrated" foods, they are fruits.
Secondly, opportunities to eat meat are most often passed up by chimpanzees in seasons when the fruits are available, in favor of the fruits, as meat is more of a fallback food that is eaten more in seasons of shortage of the more preferred fruits.
Thirdly, there is no basis in evidence for stating that meat is more valuable to or for chimpanzees than insects or nuts.
Fourthly, many individual members of a chimpanzee group will actually eat almost no meat, and the large majority eat only a very small amount (less than 5% of calories) of meat (due to their inability (i.e. in males) to catch it, or their disinclination (i.e. in females with offspring, due to the danger) to catch it, or to their inability to successfully beg for it from those who have caught it), whereas other foods are eaten in far more equal proportions among members of a group.]

(DV) [Furthermore, Wrangham fails to mention palatability and natural taste preferences, and their relationship to dietary evolution, at all… Likewise, parasites…]

(RW in PU) "…… In short, the adoption of cooking is expected to be accompanied by a series of large influences on various important biological systems, such as foraging behavior, digestive strategy, infant development, geographical range, and the regulation of social competition. …… Cooking was once considered to be an important influence on human evolution [(DV) Wrangham cites Charles Darwin, 1871] …… But in the second half of the twentieth century, such ideas largely disappeared. The main reason appears to have been the pattern that has crystallized in archaeological data.

Data shows that fire was controlled in several sites in southern Europe during the Middle Paleolithic back to at least 250,000 years ago, and probably as early as 300,000 – 500,000 years ago (James, 1989; Straus, 1989; Gamble, 1993; Monnier et al., 1994; Brace, 1995). This evidence has been widely regarded as so much stronger than any indications of the control of fire in earlier times that the Middle Paleolithic is now conveniently interpreted as the first time that humans used fire. There is admittedly scattered evidence for earlier control of fire, but none of it is sufficiently convincing to persuade the skeptics (table 16.1).

(DV) [There needs to be differentiation amongst using fire fortuitously or opportunistically obtained, controlling fire fortuitously obtained for relatively long periods of time, and *making* fire as required, which are three very different things…]

(RW in PU) "……

According to such evidence, therefore, *Homo* must have relied on raw food before the Middle Paleolithic.

In that case, if the adoption of cooking strongly influenced human biology, a suite of changes should be visible in our ancestors' evolutionary anatomy around 300,000 – 500,000 years ago.

In fact, however, the evolutionary changes in anatomy that are recorded around that time were trivial. ……

Three kinds of solution have been suggested. ……

The Late Solution.

The Late solution suggests that cooking has been adopted too recently to have had time to influence our evolutionary biology."
(DV) [Wrangham's reference to Gould's (2002) statement that speciation can occur in less than 25,000 years, is meaningless without reference to the generation time of the species referred to. Also the average time of speciation, or the distribution of speciation times, were they known, might be more relevant. Also differentiation of the time required for DNA-degenerative change, time required for removal of that, time required for selection-from-DNA-variety change, and time required for neogenesis of new function, is relevant. (see also section 3.2, p265 below)] ......
(RW in PU) "To be reconciled with the Middle Paleolithic evidence for the control of fire, this hypothesis would have to suggest that fire was controlled for a long period without leading to cooking. That solution is hard to imagine. Even wild chimpanzees take advantage of fires to eat foods that have been cooked by chance."
(DV) [Fires that occur more often in the dry season, obviously enough, that is the season of preferred food shortages for chimpanzees, when they are more hungry than usual (very hungry in fact, as juveniles cease to grow...) – and eating the cooked corpses remaining after the fire has been through hardly constitutes much of a "use" of fire, and certainly nothing resembling *making* fire....]
(DV) [Again, there needs to be differentiation amongst using fire fortuitously or opportunistically obtained, controlling fire fortuitously obtained for relatively long periods of time, and *making* fire as required, which are three very different things, and almost certainly there was a substantial period of time of evolution from the first, through the second, and to the third as a general practice...]

(RW in PU) "The Sneak Solution.

The Sneak solution accepts the idea that cooking was adopted during the Middle Paleolithic and therefore concludes that it did little to affect human evolution or biology beyond eventually causing a reduction in tooth size (beginning around 100,000 years ago, Brace, 1995). In other words, cooking "sneaked" into human culture with minimal effect. [(DV) Whoever thought up this singularly undescriptive uninformative name should try again! … Why not just the Minimal Effect Solution?...]

In its favor the Sneak solution provides a logical interpretation of the Middle Paleolithic archaeological evidence.

Furthermore, it is compatible with the idea that cooking has only trivial nutritional effects, which was suggested by Levi-Strauss (1969) and has not been completely abandoned. Against it lies the challenge of explaining why the apparently important results of adopting cooking, including a large improvement in the diet "

(DV) [What evidence is there for this, other than in palatability of some things, and amount, neither of which is equivalent to an improvement in quality?…

Statements to the effect that cooking results in a large improvement in the diet are generally not well considered.

Firstly, improving the palatability of some foods might actually decrease the quality of the diet if more suitable alternative foods were then not sought.

Secondly, changes to fiber in foods, to break it down and render it more digestible, might not be beneficial – there has been quite a lot of association of dental caries with cooked carbohydrate rich foods in the literature, as well as extracted uncooked carbohydrate – such material is more likely to stick in tooth-tooth and tooth-gum crevices than the same foods uncooked and whole.

Related to this, the increase of the speed at which the carbohydrate in a food is able to be absorbed into the bloodstream, a measure of which is the familiar 'glycemic index', by cooking, is very probably not a good thing.

Making it possible to eat a meal much more quickly than if the food were raw might not be good for digestion (a quicker and more complete digestion of carbohydrate is not the same thing as a more complete digestion of protein in the same meal, in fact it is probably somewhat incompatible with it, regards the protein food spending time in the stomach in gastric juice which is being diluted by incoming food) (see section 13.3, p392), as well as psychological satisfaction and social aspects.

The formation of 'Maillard products' in which the essential amino acid lysine is bound to a sugar and therefore not available for absorption from the digestive tract is not a good thing.
The oxidative damage of unsaturated fat subjected to high temperatures in the presence of oxygen to form peroxides and epoxides, reactive biochemicals which play a role in damage to artery surfaces and stimulation of macrophages to enter artery tissue and progress atherosclerosis (hardening of the arteries) is not a good thing.
The conversion of nuts into more allergenic foods by roasting is not a good thing.
On the other hand there is the beneficial inactivation of some toxic legume lectins, but this needs to be considered in the context of raw legumes having been reliably reported to be one of the very main foods of the large majority of the wide range of primates, allowing their ecological development as a group – is it possible that only a lesser number of legumes have harmful lectins, and that these lectins are less harmful if eaten in the context of a whole raw meal, perhaps because there are other chemical moieties in the intestinal residue that they can bind to instead of the exposed carbohydrate chains of the glycoproteins in the human cell wall membranes?
These considerations need to be made before thrusting forward one's theory as a definitive or even potentially plausible statement of the state of affairs…]
(RW in PU) " and a major change in the way in which it was distributed "
(DV) [What evidence is there for this?...],
(RW in PU) " did little or nothing to influence the course of human evolution.
A possible solution is that previous food-processing techniques (such as pounding) closely mimicked the effects of cooking (Wrangham, 2006)."
(DV) [Techniques such as pounding and grinding, would somewhat "mimic" (serve the same function, more accurately…) some effects of cooking, such as making eating the food easier, but not others, for example the effect on taste - which is not a very competent thing to omit to mention, as taste very much drives food choices…]
(DV) [Also, selection for a supralaryngeal vocal tract able to make

all of the sounds of modern humans, which was purportedly (and I have found no disagreement in the literature) not finalized in evolution until 'fully modern man' evolved 35,000-odd years ago (see section 1.11, p137 above), needs to be considered in the final reduction of human tooth size and final jaw shaping.]

(RW in PU) "The Basal Solution.

The Basal solution is the radical hypothesis that cooking was adopted around the origin of *Homo erectus* and was responsible for many of the features that characterize human evolutionary changes from australopithecines. It was proposed on the basis that many of the evolutionary changes that accompany hominization are easily explicable as responses to cooking, such as the reduced jaw and teeth, evidence of smaller gut, and yet higher energy expenditure (Wrangham et al., 1999)."
(DV) [These evolutionary changes are much more likely to be due to early tool use of digging sticks in digging for tubers, pounding rams in breaking up tubers, naturally formed hammer and anvil stones in i.e. nut cracking, naturally formed mortar and pestle stones in i.e. seed grinding, and naturally formed sharp-edged stones in tuber peeling, along with soaking of various foods.]
(DV) [And again, selection for a supralaryngeal vocal tract able to make all of the sounds of modern humans, which was not finalized in evolution until 'fully modern man' evolved 35,000-odd years ago (see section 1.11, p137 above), needs to be considered in the final reduction of human tooth size and final jaw shaping.]
(RW in PU) "This solution faces the challenge of explaining why evidence of the control of fire is scarce before about 400,000 years ago. It must also be reconciled with the traditional idea that meat eating was the prime dietary mover of the evolution of the genus *Homo*."
(DV) [What evidence and logic is there is there in support of this cherished traditional idea?... And see the previous comments just above here]

(RW in PU) "I now consider evidence relevant to each of the three solutions. ……

It is also known that a diet of raw plant food creates substantial energetic problems for humans under even the best conditions. ...... Second, even under the most favorable conditions people who attempt to restrict their diets to raw foods do not thrive. ...... In sum, the sparse [(DV) Indeed, and it must be *very* sparse!...] current evidence suggests that raw-food diets produce inadequate energy for humans even under excellent conditions, at least when the diet is dominated by plant items. ...... Third, even the most committed raw-foodists find it difficult to keep to their regime because they are consistently hungry even when they eat as much as possible."
(DV) [And later on:]
(RW in PU) "Evidence from raw-foodists indicate that under subsistence conditions humans would not survive long on the kinds of raw foods that are available in the wild."

(RW in PU) "Humans are not known to be able to survive on raw food "
(DV) [I must have died many years ago, if so – isn't it strange the way that feeling of extancy lingers on?...],
(RW in PU) " which suggests that during our evolution, we became physiologically committed to eating foods of such high quality that in most circumstances they had to be cooked."
(DV) [More arrant nonsense, see above.] ......

(RW in PU) "First, although raw plant food is evidently a poor diet for humans, a sufficient inclusion of raw meat, in theory, create an energetically adequate diet."
(DV) [Consider the basic absurdity of this statement: raw meat is about 70% water; raw nuts and oily seeds have very little water, the large majority of the weight being taken up by about 20% protein on average, and the another 70-odd% being fully fat or carbohydrate – so obviously raw nuts (600 kcal / 100 g) (or oily seeds (600 kcal / 100 g), and also cereal grains (380 kcal / 100 g) for that matter…) are more energy dense than raw meat (190 kcal / 100 g, lean; 230 kcal / 100 g, regular) - the maths here is at about a primary/ elementary school level – just open up a standard food nutrient content tables and do the elementary school maths!...
Furthermore note that high-protein diets are reported to increase the

cost of digestion by about 30% compared to high-fat diets.] ……

(RW in PU) "The *theoretical* reasons why a raw-food diet (at least when dominated by plant items even of the highest quality) is *expected* to be difficult for humans have not yet been elaborated."
(DV) [(My emphases) Consider the utterly circular reasoning in this statement!...] ……

(RW in PU) "Milton (1987, p.103) noted that human gut proportions were similar to those of Cebus and Papio, species [(DV) of monkeys] that eat many insects and process their food with their hands. She concluded that the similarity represented "similar adaptive trends in gut morphology in response to diets made up of unusually high-quality dietary items that are capable of being digested and absorbed primarily in the small intestine."
(DV) [Quite possibly – macronutrient-dense foods like nuts, oily seeds, cereal seeds, starchy tubers, fruits and insects being just as likely, if not more likely, to meet that criteria as meat…] ……

(RW in PU) "Likewise, although the human digestive system does not seem to have been compared systematically with those of carnivores "
(DV) [and what an oversight that is, at that time (2006/7)!.. though apparently plenty of writers are prepared to make vaunted (assumed) similarities a crux of their ill-formed arguments, despite the general gross anatomical appearances of the guts of humans and apes and frugivorous monkeys being very similar…],

(RW in PU) " it is clear that gut kinetics are radically different. Important, for example, although gut passage rates are similar between humans and dogs, dogs retain food in their stomachs for much longer (around 4 to 12 hours) than humans (around 1 to 2 hours; Ragir, Rosenberg and Tierno, 2000).

In addition humans show little evidence of being able to survive purely on meat diets, even when cooked (Speth and Spielmann, 1983). ……
Current data give us little hope of knowing much about the detailed

origins of cooking. For example, how was fire first controlled? How long was it used without being made? How long was it controlled without being used for cooking? And what methods of food preparation were used by humans before cooking? There are no obvious ways to answer these questions. …… we may still be uncertain whether a given population of fire users cooked their food. ……"

(DV) [Here I note the commonly reported association of cooked starchy foods (it is generally agreed, including by Wrangham, that cooking makes the starch more easy to break down, in fact breaking it down aforehand, into glucose, the free refined form of glucose being also associated with tooth decay) with dental caries (tooth decay) while the raw foods are much less so associated (Moynihan 2005).
Franck (1959) noted, while working in Africa at Albert Schweitzer's bush hospital, that African college students from either side of a sharp line of dietary demarcation differed greatly: All had massive decay of all molars and most front teeth from the area with manioc (starchy root eaten cooked as a dietary staple), peanuts, sugar cane and some meat as their diet, whereas there was hardly any decay at all in students from the area with fish as the staple food.
Similar observations were reported by dentist Weston Price (1939) notwithstanding his substantially faulty grasp of epidemiological research methodology (it was very early in the evolution of epidemiology as a science), and the multitude of imperfections in some of his reasoning on what diet might be best for humans. Unfortunately his present-day followers of the Price-Pottenger Nutrition Foundation have almost the same degree of misunderstanding as Price did back then…]

(DV) [I'll finish here by noting that Wrangham (2009) in his book Catching Fire How Cooking Made us Human, is just as fatally flawed in his arguments that we are evolutionarily genetically physiologically adapted to a diet of cooked food high in animal food.
Just briefly here I'll list the main flaws of the first chapters, leaving

the fuller detail for a book review planned to be uploaded to the Multifactor Health website before the end of November 2010, to which (as for this my writing here) he will be offered the right to full and unabridged reply to what will be an extensively destructive review of his work, as he is basically a person, Harvard Professor notwithstanding, who has some cherished ideas and beliefs about his cherished lifetime dietary habits, that he attempts to support to others and rationalize to himself with a complicated structure of selected (leaving aside the contradictory) facts and faulty reasoning, in the manner of a theologian, rather than in the manner of a scientist:

In the Introduction, Wrangham fails to note that most "hunter-gatherers" are in fact better described as "gatherer-hunters" because most of their diet is made up of plant foods;
fails to note that present-day and historical gatherer-hunters live in more marginal areas because the more choice areas were taken over by agriculture;
fails to note the concept of 'fallback foods' eaten in seasons, and very much the point here, places, where preferred foods are unavailable;
uses "butchering habilines" to describe our ancestors of two million years ago, which without any qualifier indicates that butchering was a main or the main activity of theirs, which is very unlikely from the evidence, evidence that he does not at all address in any case;
claims that fossil evidence, evidence that he does not at all address in any case, indicates the cooking of food genetically physiologically influenced our evolution from before the arrival of *Homo erectus*, which is sheer conjecture mostly disagreed with by others in the field of knowledge, and not at all able to be stated as a fact even in the least conservative reasonable assessment of the evidence.

In Chapter 1 Wrangham makes (the whole chapter is) the same utterly false claims that humans cannot readily meet their caloric (energy) requirements from raw food, a claim that I have comprehensively demolished above in this chapter section here.

In Chapters 2 and 3 Wrangham again fails to so much as even mention the alternative to cooking that is well known to all scientist/academics in the knowledge area – that food processing by the use of naturally shaped stone and wood tools, which wild chimpanzees and even some wild monkeys are well documented to do, provides the same enhanced breakdown and selective consumption of components of otherwise difficult foods.
He also fails to provide those simplest of relevant nutritional facts, that oily nuts and seeds are roughly 600 kcal / 100 gram, cereal grains roughly 380 kcal / 100 grams, and meat roughly 200 kcal / 100 grams, in talking about cooking and meat having been responsible for humans having evolved a smaller large intestine and larger small intestine, failing also to refer to the evolutionary triplication of the human salivary amylase (enzyme for breaking down starch) and starchy root vegetables and cereal grains, not to mention all the other such details of human metabolic biochemistry that I deal with {{in reviewing his Chapter 4 here (p 178-183) and in p263-86 of Vance (2010/11)}}.
And fails to address the proper chewing of food, and how most people in modern societies have become very much in the habit of not doing so, and how these relate to the digestibility of starch in raw foods.

In chapter 4 Wrangham fails to understand that humans still climb trees well enough to make nests in them as apes do;
fails to understand that with the appropriate use of sentries in a sufficiently large enough group sufficiently well armed with naturally shaped clubs and stone projectiles, fire is not necessary for survival while sleeping on the ground at a strategic location (i.e., under climbable trees, with weapons cached in them, with cover and clear ground in appropriate locations in the field of potential operations) instead of in the trees;
fails to differentiate amongst using fire fortuitously or opportunistically obtained, controlling fire fortuitously obtained for relatively long periods of time, and *making* fire as required;
fails to address what proportions of archaeological sites at the different archeological times had evidence of fires and did not have evidence of fires; ….

And the flaws go on and on, page by page, which I will detail in a separate book review as noted.

Approximately the only useful information to be derived from (the first four chapters of) this dreadfully flawed book is that cooking probably does increase the digestibility of starchy foods, and that cooking has had a psychological, social and cultural effect on human evolution, but the lay reader will be put to much effort to disentangle these facts (available better dealt with elsewhere) from the nonsense, and the cognate reader either much annoyed at the faults or pleased to have themselves a better understanding of the matter than Wrangham, probably both at once…

And how amazing that all those in the long list of prestigiously placed people quoted in praise of the book in the first four pages inside its cover should have such little understanding of the topic, yet are so forthcoming in voicing their effusive (and probably reciprocated somewhere down the line or in the network to mutual personal financial if not intellectual benefit?...) praise for the book…

Again I quote, "It is better to remain silent and be thought a fool, than to speak and remove all doubt."…]

Now follows the email dialogue between Prof. Wrangham and myself, regarding section 1.14 of my book as reproduced immediately above, in response to my emailing him (as I did the other authors in Ungar (2007)) a copy of my book "The natural Human Diet? The Optimal Human Diet? The known and the unknown, the possible and the probable – an analysis of the evidence" together with my offer of right of reply to my critical reviewing of his work and undertaking to attach at the end of the book any responses complete no matter how negative (the same offer is made below regarding my work here in the fully detailed critical reviewing of the whole of Prof. Wrangham's book Catching Fire done here, and any responses made to, and dialogue ensuing from, this offer will be attached below as Appendix 2 on the

Multifactor Health website copy of this book, none having been forthcoming as yet):

The following was received from Professor Richard Wrangham of Harvard University, on 12 November 2010, other than my words in response in square brackets (DV) [Richard, abc … xyz.] being interspersed among his words now here in quotation marks (RW) "abc … … xyz" and the resultant below being sent back to Prof Wrangham. Corrections of my own words made after the facts of the communications are identifiable as "(correction: abc … xyz.)" – it can readily be seen that none of these would alter the substance of the discussion much, although the omission (in haste, not due to lack of knowledge, as my book itself would make clear) of legumes from the food types is something I should very much have been corrected on (but wasn't)… :

(RW) "Dear David Vance,
Thank you for your courtesy in inviting a reply to your in-depth analysis. You have certainly read a lot of material!
I note that on pages 43, 170 and 385 you refer to my "statements here
and in his book Catching Fire How Cooking Made us Human (2009), that
humans cannot meet their caloric requirements from raw plant foods",
and you propose that I am wrong. This raises two points.
(1) I don't remember saying that "humans cannot meet their caloric requirements from raw plant foods", but maybe you can find somewhere I
did so. If so, I apologize for my mistake. Actually I know perfectly well that people can survive and thrive on a raw diet, provided that it is a modern urban context using the global food resource,
domesticated foods, etc. What I remember saying is that (a) people who
eat 100% raw foods often do not meet their caloric requirements; (b)
no humans are known to have lived successfully off 100% raw

foods in
the wild; and (c) based on the available evidence so far, humans would
not be able to live successfully off 100% raw foods in the wild. I would be most interested if you have evidence against these points."

(DV) [Richard, in chapter 16 of Ungar (Ed, 2007) you say the following:
"(p315 lines 1-2) It is also known that a diet of raw plant food creates substantial energetic problems for humans under even the best conditions. ; (lines 25-26) Second, even under the most favorable conditions people who attempt to restrict their diets to raw food do not thrive. ; (41-3) In sum, the sparse current evidence suggests that raw-food diets produce inadequate energy for humans even under excellent conditions, at least when the diet is dominated by plant items. (p316 lines 1-3) Third, even the most committed raw-foodists find it difficult to keep to their regime because they are consistently hungry even when they eat as much as possible." All of these exact quotations of you appear on p166 of my book.
"(p316 line 15) Humans are not known to be able to survive on raw food, which suggests that …" which exact quotation of you appears on p171 of my book.
Furthermore, chapter 1 (Quest for Raw-Foodists) of your (2009) book, Catching Fire How Cooking Made us Human (which book I will deal with shortly in a separate review), is devoted to attempting to establish just this fallacy, which very much constitutes one of the crucial bases for your theory.
Furthermore in noting failure to meet caloric requirements it is necessary to different (correction, differentiate) cases of anorexia nervosa from the whole group, or the analysis will be fatally flawed.]

(RW) "You very sensibly say that a diet based around oily nuts could provide
adequate calories. But you doubtless realize that no wild habitat has been found where such a food source is available for sufficient duration to sustain people. It would be lovely to find one, and the

question would then be whether such a habitat would ever have been
sufficiently widespread to have evolutionary significance."

(DV) [Richard, I did not say that a diet "based" on or "around" oily nuts could provide adequate calories – I have made it abundantly clear in my book that a diet including oily nuts and oily (and other) seeds and cereal grains and legumes and starchy root vegetables and fruit, even allowing for a good serve of green leaf vegetables alongside, would provide adequate calories – and that, obviously enough, the proportions of these in meals and in the diet would vary, including by season – and as for other animals, seasonal surfeits and deficits apply.
As to what habitats have been found providing a sufficiency of these, you fail to note that the point is not the lack of habitats found, the point is the lack of any attempts to trial a (sensibly-proportioned, unlike some of the very defective studies you've cited in chapter 1 of your (2009) book…) raw-food diet of wild foods in any habitat. Lack of evidence is not the same thing as evidence of a lack. Gatherer-hunters use fire for many reasons, not the least of which is for technological/skill/magic social status, among flavor and temperature reasons.
Also, the knowledge of how to utilize the wild foods available is obviously not to be assumed in setting up any study, as selecting appropriately from among a complex variety of foods is obviously the reason the orangutan has such a large brain, as it obviously doesn't have such a large brain for social interaction – and the wild human/hominin situation would be likely to have even more dietary complexity, including the use of tools from the earliest of times.
Furthermore, you fail to note that the large majority if not all of the richer habitats have been taken over by agriculture, leaving only more marginal areas in a wild state.]

(RW) "(2) The repetition on pp 43, 170 and 385 suggests that your book is
still in rather an early stage. Would it be more sensible to invite comments when you are nearer completion?"

(DV) [Richard, the book is in exactly the form that I want it. Repetition of crucial points in appropriate contexts is not at all a bad thing (and at times unavoidable, as something may be relevant to a plurality of topics), and where you find it, it is quite deliberate.]

(RW) "Good luck with your project."

(DV) [Richard, thank you, and good luck with the evolution of your own knowledge in the area of human diet, and I hope I may be, by way of my book, of some assistance with that.]

(RW) "Richard Wrangham"

The following response to the above communication was received from Professor Richard Wrangham of Harvard University, on 16 November 2010, other than my words in response in square brackets (DV) [Richard, abc … xyz.] being interspersed among his words now here in quotation marks (RW) "abc … … xyz" and the resultant below being sent back to Prof Wrangham:

(DV) [Richard, in response to your communication below, as before, my responses are interspersed in your communication in square brackets, as this sentence is.]

(RW) "Comments re David Vance

Email, Nov 15 2011

"Richard, in chapter 16 of Ungar (Ed, 2007) you say the following: "(p315 lines 1-2) It is also known that a diet of raw plant food creates substantial energetic problems for humans under even the best conditions. ; (lines 25-26) Second, even under the most favorable conditions people who attempt to restrict their diets to raw food do not thrive. ; (41-3) In sum, the sparse current evidence suggests that raw-food diets produce inadequate energy for humans even under excellent conditions, at least when the diet is dominated by plant items. (p316 lines 1-3) Third, even the most committed raw-foodists find it difficult to keep to their regime because they are consistently hungry even when they eat as much as possible." All of these exact quotations of you appear on p166 of my book. "(p316 line 15) Humans are not known to be able to survive on raw food, which suggests that ..." which exact quotation of you appears on p171 of my book."

Thanks for finding these. In case there is any misunderstanding, I hope you appreciate that my statements (which as you will know are based primarily on the Giessen raw-food study, albeit supplemented by every other study I found) refer to populations. You sound to be a very physically fit individual who does well on a raw diet, and you write almost as if you are personally affronted by the idea that people do not do well on raw food diets in general!"

(DV) [Richard, if there is a tone of affront it is less personal and more due to:
The affront felt by a scientist committed to the fullest rigor in science when the appropriate degree of rigor has obviously not been applied to producing a work of science, particularly when that work is by someone who holds an academic position the prestige of which is not backed up by the quality of the work, and particularly when the author manifests assumptions of having a superior grasp of a topic on which it seems likely to me that my own understanding is in fact superior, my lack of a prestigious (or any, as yet...) academic position notwithstanding;
The affront felt by someone who sees publication of poorer work than their own, achieved not by scientific merit, but rather by network connections, when their own work goes unpublished due to not being part of such networking, particularly if it points out important flaws in the networkers' work;
The affront felt by a worker in health science working towards the reduction of the suffering of those beings (of whatever species) on earth who are suffering the most, to the waxing lyrical about dietary habits (born far more of cultural evolution than physiological evolution), the pandering to which via the cage-farming of animals is one of the very greatest and most easily

preventable sources of suffering on the planet, couched in pseudoscience born of culture not of reason;]

(RW) "At any rate, please make clear in your book, as I hope is clear from a full reading of the passages that you cite, that I was summarizing the data on average."

(DV) [Richard, no that isn't clear at all – all of your statements there, in the full context of all of what surrounds them, are clearly made in the absolute – there are no references to exceptions or averages – and clearly you are very wrong, and as a scientist should admit that you are wrong, and very wrong, because it is a matter crucial to your theory…]

(RW) "(To avoid the kind of problem that you have drawn attention to, perhaps I should have written "It is also known that *on average* a diet of raw plant food creates substantial energetic problems for humans under even the best conditions." And so on for the other lines.)"

(DV) [Richard, no, that would not do either – if one is talking about the best conditions, the only way that could be sensibly interpreted is that the diet is suitably composed with respect to proportions made up by the different food groups: nuts, oily seeds, cereal grains, other seeds, starchy root vegetables, (calorific) fruit, and green leaf vegetables – any other meaning would be plainly silly and irrelevant – you were wrong, and should admit you were wrong…]

(RW) "Your general tone is that my conclusions were not merely wrong but somehow biased and/or corrupt."

(DV) [Richard, it does indeed seem to me that, like many other scientists who apparently have not a good grasp of psychology, such as might be applied to understanding their own mental processes regarding their cherished enculturated habits better, you are indeed biased by your cherished psychosocially and culturally programmed habits of carnivory and eating cooked food. I note you haven't even got as far, in all your years of work as a primatologist, and now venturing into human nutritional science (a competency background of mine…), as to simply try *yourself* various *sensible* formulations of raw food diet…]

(RW) "But do you know of any studies that contradict the findings of the Giessen study? I did not you citing any such research. Sorry if I missed it – please send if possible."

(DV) [Yes, my own personal 20-year experiment, as detailed in chapter 10 of my book. Sorry if you missed it… Do you think it would achieve publication in the 'peer-reviewed' scientific literature without me being part of some little network?....]

(RW) " "Furthermore in noting failure to meet caloric requirements it is necessary to differentiate cases of anorexia nervosa from the whole group, or the analysis will be fatally flawed."
If a significant proportion of the subjects in the Giessen study were anorexic, yes, that would be a problem, I agree. The care that the Giessen researchers took, and the consistency of their results (and other studies) in finding caloric shortages among raw-foodists would make a hidden role for anorexia surprising, but it is theoretically conceivable."
(DV) [Richard, having looked at the journal articles (Koebnick et al 2005 and the abstract for 1999) and a website by the authors, it is obvious that your ability to make authoritative comments on the "care that the Giessen researchers took", doubtless due to your lack of epidemiological and population research methodology background and competence, is none too good – there is no mention of the Giessen researchers screening out anorexia nervosa patients, and in fact in the population there is a continuum from milder (under-)eating disorder cases through to more severe ones, and the Giessen methodology for acquiring their subjects would have not at all removed any other than the most severe staggering cases of (under-)eating disorder from the subject group – furthermore, the association with underweight is further confounded by the fact that roughly 50% of the subjects took up the raw food diet as a way of dealing with a pre-existing disease.
Furthermore, regards the Evo Diet Experiment, as mentioned in

your (2009) book, it is the most basic of knowledge that salt retention (the need for retention in the body of more water to hold excess table salt (NaCl, from the usual western diet) in solution before it can be excreted, actually) would account for maybe most of the loss of body weight when it was no longer necessary on the Evo Diet – failure to address this aspect indicates a serious lack of competence in the science of physiology.
Furthermore, consistency of results means nothing if all are making the same mistakes, which is not a very rare event, actually…]

(RW) "On the broader topic of the reliability of the information on the effects of a raw diet I take your general point that this area is not very studied yet. I completely agree with you that much remains to be learned. In my attempt to review this area from as many directions as possible I found what I considered some remarkable evidence all pointing in the same direction, but we still need a prospective study in which individuals start on a cooked diet and are then medically monitored as they eat increasing amounts of raw food in their diet. In my lab we are thinking about trying to design such a study in my lab, but it is a major undertaking – and my main work is as a primatologist. Hopefully the ideas presented in *Catching Fire* may lead others to undertake such studies even if my own group cannot find the resources."

(DV) [Richard, it's actually a very *easy* and *in*expensive undertaking – just do it yourselves! Gain the basic competence in nutrition necessary to formulate the appropriate composition by food groups of the diet, and eat the diet yourselves! If you are unable to gain that basic competence in nutrition, I have it and am available to help you with that, for a modest fee and the accordance of authorship in the doubtlessly resultant journal article…]

(RW) " "the point is the lack of any attempts to trial a … raw-food diet of wild foods in any habitat."
Yes! I greatly look forward to such a study (and have tried to encourage some efforts in that direction).

"selecting appropriately from among a complex variety of foods is

obviously the reason the orangutan has such a large brain"
You would find a lot of dispute on that point among the people who study primate brain evolution. It is only one of numerous possibilities; nothing is "obvious" yet about that topic."

(DV) [Richard, yes, it is possible that the orangutan's large brain is a holdover developed for e.g. social cognition in an ancestral social primate, but note that the gorilla's social complexity is not very high either – and in any case failure to address these aspects, as is common enough in the literature, is indeed a major failure – in the literature what I find mainly is the assumption that the large primate brain is due to higher social function, without any reference to or explanation for the nonconformity to this of both the orangutan and the gorilla…]

(RW) " "Furthermore, you fail to note that the large majority if not all of the richer habitats have been taken over by agriculture, leaving only more marginal areas in a wild state."
This is an interesting topic. I agree with you about most hunter-gatherer habitats being marginal. (But not all: e.g. those in the Okavango in Botswana, Asmat area of New Guinea, and Murray-Darling basin of southeast Australia were high-quality in many ways.) But the problem that I come up against is that great habitats like the Nile Valley or the Ganges Basin offer plant resources that mostly need to be cooked. What is your idea for an optimal pre-agricultural habitat where humans could have lived raw?"

(DV) [Richard, somewhere with relative plenty of some suitable combination of nuts, oily seeds, cereal grains, other seeds, starchy root vegetables, fruits and green leaf vegetables – when I say relative, I mean in the context of our prehistory – where such a criterion might be met today I do not know, as I am far from expert in botany, ecology and geology.]

(RW) "P 219-220 Hunting by chimpanzees a response to food shortage? Actually the evidence mostly points in the opposite direction. See Gilby & Wrangham 2007, attached."

(DV) [Richard, it is quite consistent among the literature I've reviewed in my book that chimpanzees hunt mostly when preferred food is in short supply.

In any case, in the article of Gilby and yours that you've cited just here, after controlling for the effect of the number of adult males in the group being observed, the odds ratio for hunting rather than not hunting on encounter of red colobus monkeys, although statistically significant (not very, p = .013) due to the number of encounters being large at 790, was only 1.01, that is, a 1% increase! Big deal!.. Furthermore, the actual success rate of the hunting, as a measure of the actual commitment of the hunt, rather than it maybe just being activity for excitement, with time and energy on one's hands (like young male elephants hassling rhinos for no apparently good reason other than practicing hassling, or humans ditto…) needs to be a focus of analysi(e)s. Also colobus characteristics such as tree of occupation characteristics, area of occupation relative to area of chimpanzee occupation, age (regarding dependant young) distribution, pregnancy, vocalization, and nutritional state.

Furthermore, as someone perhaps better qualified than yourself in such matters, I state that when you write an abstract for such a journal article, you need to include the quantification of the effect (association) size (in this case the measly 1%, but the 95% confidence interval for the odds ratio (which needs to be provided, but you didn't…) would very likely have included a lower value of less than 1.005, that is, an increase of a tiny 0.5%...), not only its statistical significance – with large enough numbers even the most *virtually in*significant effects (associations) are *statistically* significant.

Now, in that work of Gilby and yours you stated a U-shaped relationship between nutritional abundance and hunting activity was indicated – why do you not state so here – all you speak of here is one end of that curve – which in any case, once the tiny effect size is known, would not be a U shape, but the only (correction: 'only the') first half of the U, finishing in a tiny rise so utterly trivial that the naked eye would not even be able to see it…. And accordingly Fig 1 in your work is a gross misrepresentation of the actual results – by eye, one would not be able to see *any* rise at the right hand side of the probability line, because what rise there is,

is so tiny – and by the way, a probability of 1 means that the event *always* happens – your representation of an odds ratio of 1.01, a 1% increase in probability, as a 100% certainty (probability = 1) is wrong to the point of being bizarre – which I would expect anyone who had passed an introductory course in statistics at university to pick up in very short order...
These certainly look like misrepresentations for the sake of refusing to admit your mistakes, both here in this dialogue, and also in your theorizing on the subject of the larger discussion as addressed in your and my books.]

(RW) "NB In various sites, such as Kibale, chimps eat almost no insects – we have excellent data on that point after 20+ years of observation."

(DV) [Richard, for a start, what eats wild figs, eats insects, incidentally. And ditto for other wild plant matter, particularly fruit. I'm astounded that you would attempt to argue this. You may have a lot of data, but its excellence in this respect is obviously far from established...]

(RW) "p. 220 "suggestive of predation not being naturally a strong primary drive." That is a difficult concept. Everyone who studies chimpanzees is impressed by their interest in meat."

(DV) [Richard, from your own (2009) book Catching Fire (p142): "Consider chimpanzees, who hunt little and whose raw-food diet can safely be assumed to be similar to the diet of australopithecines [This latter claim is very defective in its failure to consider differences in grass seed and starchy root vegetable use, which are far from trivial matters...]. At Ngogo, Uganda, chimpanzees hunt intensely compared to other chimpanzee populations, yet males still average less than three minutes per day hunting."]

(RW) "True, the Gombe studies can be criticized for the banana period. But we now have lots of data from provisioned and unprovisioned sites. Not only do chimps like to hunt (when they have the energy to do so) but also, once meat has been captured, the intensity of interest is huge."

(DV) [Richard, interest is intense because the lean seasons, in which most hunting does take place, are so lean that growing juveniles actually cease to grow – interest is intense because hunger is intense…]

(RW) "Read Goodall's (1986) book, which captures the interest in meat for any population."

(DV) [Richard, ditto.]

(RW) "Remember, females and young will work hard to just get a few drops of blood as they fall to the leaves on the forest floor!"

(DV) [Richard, ditto.]

(WR) "It is interesting that chimps don't prey on some animals that they might be expected to, such as bats in weaver nests and maybe monitor lizards. But with regard to fish, remember that chimps definitely are averse to eating animals that they find dead; and catching fish live would be very difficult. Chimps do eat lots of other things – the list of species is somewhere between 50 and 100 by now if I remember right."

(DV) [Richard, as noted in my book p221 citing van Lawick-Goodall (1968), whose work you must be aware of, that chimpanzees have been well tested to show very little interest in fresh carcasses of usual prey animals placed trailside, while they are very interested in fruits offered similarly, surely must suggest that meat is not a preferred food?...]

(RW) "p. 220 "Wrangham's more detailed picture (1975) based on 82 of the most reliable predation records in the Gombe Stream Research Centre files, covering four different periods [we need to know the total time to consider the correctness of the calculation!...We also need to be able consider any trend across the four different periods!...]

I am not sure what you are referring to here. I did calculate predation rates with respect to time, and everyone else does so too, as you can see in Stanford, Mitani and Watts etc."

(DV) [Richard, what I am referring to here, which would be obvious enough, I would hope, to anyone with sufficient a background in research methodology and data analysis to carry out an appropriate such work, is that one needs to know the representativeness of the periods from which the data that was analyzed came – so one needs to know which months of which years the data came from – and also, one needs to be able to compare the four different periods from which data came with each other – otherwise very important data detail is missing/ hidden/ obscured…]

(RW) "p. 163
"It should likewise cause a substantial drop in the amount of time that individuals spend
chewing, with large consequences for the species activity budgets (Wrangham and
Conklin-Brittain, 2003).
[But cooking also requires time; and the smoke may attract unwanted attention that then
has to be guarded against; and the food being cooked is less immediately portable than it
was before should exiting the area become necessary or strategic; and food can be
processed for young children using naturally shaped stone and wood tools…]"
The only thing that I disagree with here is your use of the word 'But'. You could equally well have said 'And as Wrangham expressly notes, cooking also requires time (etc)…' Something we agree on!"

(DV) [Richard, no, I think the correct word is "but", because your implication, from the context of the surrounding text, is not one of merely a rearrangement/reallocation of the energy budget, which I would roughly agree with, but a saving in the energy budget, which is far from obvious or reasonably assumable…]

(RW) "p. 163 "[Firstly it needs to be noted that the most valuable, and valued (sought after) foods for chimpanzees are not "concentrated" foods, they are fruits."
Not really. When chimpanzees hear evidence of a predation (or see red colobus in a vulnerable position) >95% of the time they leave their fruit-eating to go and try to get meat. On the occasions when they don't, there are rather obvious reasons for it (e.g. the chimp is too timid.)"

(DV) [Richard, again, from your own (2009) book Catching Fire (p142): "Consider chimpanzees, who hunt little and whose raw-food diet can safely be assumed to be similar to the diet of australopithecines [This latter claim is very defective in its failure to consider differences in grass seed and starchy root vegetable use, which are far from trivial matters…]. At Ngogo, Uganda, chimpanzees hunt intensely compared to other chimpanzee populations, yet males still average less than three minutes per day hunting."
Now, are you trying to say that chimpanzees know where to look for fruits but are totally clueless about where they might find colobus by either looking or listening?...]

(RW) "p. 163 "Fourthly, many individual members of a chimpanzee group will actually eat almost no meat, and the large majority eat only a very small amount (less than 5% of calories) of meat (due to their inability (i.e. in males) to catch it, or their disinclination (i.e. in females with offspring, due to the danger) to catch it, or to their inability to successfully beg for it from those who have caught it), whereas other foods are eaten in far more equal proportions among members of a group."
That's all roughly correct. But it is not an argument against meat being valuable."

(DV) [Richard, for the Galapagos Islands tortoises (normally virtually totally herbivorous) that ate their dead fellows and so were saved the same fate of starvation because all the grass had been eaten, that meat was valuable – that is not the same thing as it being either preferred or optimal… and the same applies to apes, and to humans, if you could but consider it aside from the cherished dietary habits that you have been habituated, socialized and enculturated into (please excuse the element of technical redundancy here, but these are very important concepts in understanding why we do what we do…)]

(RW) "p. 173 "In Chapters 2 and 3 Wrangham again fails to so much as even mention the alternative to cooking that is well known to all scientist/academics in the knowledge area – that food processing by the use of naturally shaped stone and wood tools, which wild chimpanzees and even some wild monkeys are well documented to do, provides the same enhanced breakdown and selective consumption of components of otherwise difficult foods."
(1) True, I say little about this topic in Chs 2-3, but I do mention it (p. 49 in Ch 2).
(2) Surely it is important to distinguish between tools used to get access to food (which is what chimps and capuchins do) and tools used to physically process food (which I think is unknown in wild primates)."

(DV) [Richard, surely the difference between cracking open a nut and belting something to crush it (which latter I seem to recall some ape populations and some monkeys have been recorded as doing also) is largely a matter of semantics in "accessing" vs. "processing"? For the purposes of the larger questions, it's all much of a continuum – surely it's a very small evolutionary step for an australopithecine to go from cracking nuts to grinding seeds and mashing tubers, and an obviously very plausible one?...]

(RW) "(3) You refer to "the same enhanced breakdown and selective consumption of components of otherwise difficult foods". But this is a guess: very little work has been conducted to assess the relative amount of 'breakdown' of food due to non-thermal vs. thermal processing, so you cannot legitimately claim "the same enhanced breakdown". Rachel Carmody has been doing some new research on this topic at Harvard, and will shortly publish results that you should be interested in."

(DV) [Richard, on p79 of your 2009 book Catching Fire, with regards to pythons eating meat you state that cooking and grinding had equal (12.3%, 12.7%) effects on increasing digestibility, the effects being additive – how this relates to humans and our different foods is in question, but it is very relevant that humans are very obviously designed to chew, whereas pythons absolutely do not chew at all… In any case, I personally know the difference between defecating lumps of improperly/un-chewed starchy root vegetables and passing properly-chewed stuff that certainly looks like more digestion has taken place, and I have experienced the amount of energy available to me from properly chewed potatoes and sweet potatoes – much the point is modern people being habituated to not chewing food properly…]

(RW) "p. 172 "Wrangham (2009) in his book Catching Fire How Cooking
Made us Human, is just as fatally flawed in his arguments that we are evolutionarily
genetically physiologically adapted to a diet of cooked food high in animal food" This puzzles me. Where did I say that we are "genetically physiologically adapted to a diet … high in animal food"? Are you mixing me up with Stanford?"

(DV) [Richard, I wrote the meaning, not a quotation – if it sounds a little clumsy due to redundancy, bear in mind that as noted I wrote my book with the average person in mind, so some things will be redundant for some readers, while helping others to understand. If that's not what you were referring to, your writings certainly give me the impression that you think we are by evolution physiologically adapted to a diet high in animal food – have I formed or received the wrong impression?...]

(RW) "p. 237 "In any case it is quite evident that the proportion of chimp diet, at this location, constituted by vertebrates has not been at all properly quantified – and what
quantification that can be readily done from the data, as here,

suggests that the proportion
of chimp diet (in this location) constituted by vertebrates of any size taken together is
very low, and vary variable according to environmental conditions."
That's right. In Kanyawara I did not feel ready after 3 years of data (with chimpanzees that were still in the process of being habituated) to make a calculation of the amounts eaten, since there could have been many sources of error. But now that we watch our chimps for around 3000 hours per year I can be confident that meat-eating in Kanyawara is much less than in Ngogo, Gombe or Tai. The reason is probably partly because our community has fewer males, and partly because they have a forest that is less productive of fruit."

(DV) [Richard, the last sentence here contradicts your earlier assertion that a plenitude of fruit gives rise to increased hunting…]

(RW) "The role of particular individuals could be important too. See Gilby et al 2008."

(DV) [Richard, agreed, and I guess you'll agree that the Pumphouse Gang baboon study on this aspect will remain a classic of primatology forever – notwithstanding other work I've not yet read…]

(RW) "p. 115 "Instead Wrangham and colleagues claim that a dietary change to increased tuber
consumption
[It seems to me much more likely that tuber consumption had been increasing and/or
substantial for a long time before the evolution of H. erectus – the early hominin can see
food acquisition of tubers by other animals, and it can readily evolve the use of the
digging stick to facilitate their acquisition for itself….]
enabled the evolution to H. erectus."
That is not what we intended. Our idea was that USOs were already

a part of the diet, but that cooking – of both USOs and meat – enabled the evolution of *Homo erectus*. As various colleagues and I have argued elsewhere, USOs were likely critical in the diets of pre-humans."

(DV) [Richard, so Bunn got that (among many other things…)(conveniently, for his theorizings?) wrong, misrepresenting you – and we have the correction from you now. And this is an example of just what I've avoided in my book by using direct quotations of what people actually did say, rather than rephrasing it to something else…]

(RW) "p. 222
"Excluding the third method, which by definition is always successful, Gombe
chimpanzees share meat at an average rate of once every 8.6 minutes.
Requests, which involve active solicitation with postures, gestures and vocalizations,
occur every 6.5 minutes and are successful every 22.5 minutes (Teleki, 1975)."
Much has been discovered since Teleki's brief study. For example Gilby, I. C. 2006. Meat sharing among the Gombe chimpanzees: harassment and reciprocal exchange. Animal Behaviour, 71,953-963."

(DV) [Richard, I'll have to read that later – the harassment I could hardly doubt, particularly given the hunger of the lean season when most hunting occurs…]

(RW) "p. 164
[Wrangham's reference to Gould's (2002) statement that speciation can occur in less than 25,000 years, is meaningless without reference to the generation time of the species
referred to. Also the average time of speciation, or the distribution of speciation times,
were they known, might be more relevant. Also differentiation of the time required for

DNA-degenerative change, time required for removal of that, time required for selection-
from-DNA-variety change, and time required for neogenesis of new function, is relevant.
(see also section 3.2, p265 below)]
Generation time does matter, fair enough. But the fact remains that evolution can be very fast. I give a couple of examples in *Catching Fire*."

(DV) [Richard, fine, but those differing DNA mechanisms are crucial to understand, in their differences – as noted, I'll be providing a detailed critical review of your book Catching Fire in the near future.]

(RW) "p. 198
"But except in the most extreme environments, edible plants are abundant and available in sufficient variety to allow a wide range of choices. From fruits, nuts, beans, roots, tubers, stalks, bulbs, berries, melons, and gums, leaves, fungi and flowers a highly diversified, nutritionally rich, and often extremely tasty diet can be obtained.
[And I agree, whereas Wrangham (2007, see section 1.14 here above) does not…]"
That's right: I clearly disagree with the notion of "nutritionally rich", unless that means diverse with respect to micronutrients. My experience tells me that the notion of a ready supply of food is derived from the industrialized and developed world. It does not apply to wild primates (and other animals), for which we have abundant evidence of extra food causing a rise in population growth rates and population size, and numerous others kinds of evidence showing that they live on the energetic edge. It does not apply to hunter-gatherers, for whom similar kinds of data abound. Nor does it apply to much of the third world, where people routinely go to bed hungry every night."

(DV) [Richard, I come back a little from my stance on this, but only insofar as I think that there would have been an average sufficiency such as human population numbers/density were probably able to be very roughly what they are/were for other apes – yes, I agree that all wild primates, and many other animals, reproduce such as to increase in number up to the carrying capacity, interacting with the level of predation by cats etc., and disease (don't forget these), where they live, roughly, as you put it, "on the energetic edge."]

(RW) "p. 18
"Humans, and therefore all of our evolutionary ancestors since the common ancestor
of humans and chimpanzees, still climb trees quite well, just not as well as the apes do -
certainly modern humans and all of our hominin ancestors climbed trees well enough to
be harvesting and eating fruit from them as a daily staple."
If you mean that fit young males could climb up and throw fruits or branches down, I would agree. But if you suggest that humans fed like apes, i.e. each for themselves, which would mean that babies, old men, middle-aged mothers etc were required to climb for themselves, your proposition seems unreasonable. You are doubtless a very fit and able male, but think of others! Even nowadays, the major source of mortality for some hunter-gatherer men is falling out of trees, and they do not perform the kind of climbing that chimpanzees do. Among the Efe pygmies that I worked with, they stay close to the trunk, often use vine ropes, etc. And old men get young men to go up on their behalf."

(DV) [Richard, I'm a genetically physically average 49 year old male – old enough to be a grandfather, old enough to be more or less redundant for the survival of my progeny, in the prehistorical and evolutionarily relevant contexts, and very few wild animals (and I assume this applied to our evolutionary ancestors also) die of pure old age – also I carry permanent lumbar disk and hamstring tear & adhesion injuries – I have known many genetically physically average enough 60 and 70 year olds who could climb trees well enough to harvest fruit - and the use of a forked stick for pulling in fruit/branches would have been available and almost certainly used by even the earliest hominins – and in those days, hominins were either fit, or maimed; no-one got unfit simply through lack of physical exercise, as is often the case these days… And yes, I don't doubt there were various divisions of labour such as to reduce risk, and this would include the riskier climbing, which is far from all climbing for harvest… If the Efe pygmies couldn't climb better than me, there must have been something the matter with them, or with the observation of them, because I've seen film footage of people whose climbing ability makes mine look pretty basic, although mine is much better than the average westerner, simply because I've actually done some…]

(RW) "p. 173
"In chapter 4 Wrangham fails to understand that humans still climb trees well enough to
make nests in them as apes do"
Same kind of comment. I have on occasion reached a chimpanzee nest and lain in it. But I cannot imagine being able to construct one. The balancing alone would be very difficult, and the strength required to bend and break branches would be beyond me, and I think would have been beyond me even when I was at my fittest."
(DV) [Richard, one needs to know what you were fit for – if it was for walking all day, that is not relevant to branch-breaking – if it was for climbing trees for hours of every day, then your upper body strength would have been very much greater – it might take longer but I'm pretty sure I could make a nest, in the right type of tree, and it would be strength, not balance, that would limit me – notably, it seems you've not yet tried to make a nest yourself, even after all these years of primatology…]

(RW) "You list an additional series of "flaws" here. To pick up on a couple… (1) I agree that it is fun to think about being able to sleep on the ground without fire. But remember that many nights are moon-less. Are you really prepared to try relying on sentries and access to climbable trees on a pitch-black night with lions roaring close by? And again, this is not just for fit males. The old, the young and the weak have to be able to be safe too!"

(DV) [Richard, for survival of the species, in wild nature, the weak

do not have to be safe – in fact they are usually eliminated by one circumstance or another, and this very often includes the aged.
If the lions are roaring, I (hominid) won't need the sentries to wake me to climb the tree.
In a big enough group (and we (hominids) are in larger groups, because we are smaller and less well armed, than recent gatherer-hunters), which is much the point in considering the survival methods of our earliest hominid ancestors, yes, we rely on multiple sentries, to wake us appropriately, and then we defend as a group (with our weapons of throwing stones, clubs and pointed sticks/ primitive spears) such as to allow the mothers and children to get up the tree(s) first. Or we are sleeping on a barely-accessible ledge, as the gelada baboons do for safety.
Maybe on pitch-black nights we slept in the trees, or spelled sentries more often so they could be sharper in their vigilance, and posted more sentries, or had more people awake more often.]

(RW) "(2) I pointed out that fire does not have to be made when first used (see CF p. 191: "The steps to managing fire need not have involved the difficult process of deliberately making it." (3) The question of the proportions of archaeological sites that do and do not have evidence of fires is really difficult to answer. One person's evidence does not satisfy another; some sites are preserved in such a way that it is unreasonable to expect evidence; and the number of sites in any one time-depth is very few by the time you reach the Lower Paleolithic."

(DV) [Richard, nevertheless the quantification, including with 'sensitivity analyses' that address levels of evidence, needs to be made in order to have a meaningful discussion about the matter.]

(RW) "Of course you are right that I am challenging received wisdom about the antiquity of fire, and many such as yourself think I am wrong."

(DV) [Richard, surely by now you must know that I am less a follower of received wisdom, and more an analyzer of evidence…]

(RW) "But at least the biological evidence that I have assembled is causing some archaeologists to look again at the evidence."

(DV) [Richard, the biological evidence that you have "assembled" is a human digestive tract suited for a relatively smaller volume of food than the digestive tracts of the apes – as I think I've substantially demonstrated in my book, this is more likely to have been facilitated by a diet composed of suitable and readily enough available proportions of nuts, oily seeds, cereal grains, other seeds, (correction: and legumes) starchy root vegetables and fruit, along with some amount of leafy green vegetables, with the use of primitive tools for digging, cracking, grinding and mashing, and drying for storage and rehydration via soaking as required, than to have been facilitated by the use of fire for cooking.]

(RW) "It is not an open-and-shut case, as most archaeologists agree. The exciting scientific approach is to stay open-minded and see what the next decade of evidence brings. If a boundary period can be found before which the evidence for control of fire is satisfyingly absent, and after which it is strongly present, our knowledge of the past will be improved. But during the time since I began using biology to attempt to reconstruct the past, the strong evidence for the use of fire has gone from 400,000 ya to almost 800,000 ya. Who knows what the future might bring?"

(DV) [Richard, firstly, a very small number of sites indicates, much less proves, little about the general state of fire use in the periods of interest/ debate. Quantification, and identification of possibly relevant site discovery and preservation biases are necessary…
It must be borne in mind that fire that is merely controlled (not made) is at risk of being frequently extinguishing by rain, and in some wet seasons in some locations would only have been able to be maintained by storing dry timber/fuel in a particular place(s), near to another particular place where the fire could be maintained – regardless of where these two types of sites were relative to the actual place of the use of the fire for living, it seems that some degree of sedentariness or localization would be necessary to be able to ensure the maintenance of the fire in this way – in which

such cases fires used for living by hominids should have been recurrent enough at locations for evidence to potentially be findable. Otherwise, after the rain, another lightning strike or volcano eruption (volcanoes might have been relatively permanently available in some areas) must be waited for to start the next fire to be captured and maintained.]

(RW) "That is my general approach. This is a fascinating area, and I am open to correction. Maybe I am wrong that cooking strongly tends to increase the amount of energy that we obtain from our food. But that is the way that the evidence points right now. If experiments can be done to show I am wrong, fair enough! That's science. But until then, the energy theory of cooking seems to me a really interesting way to re-thinking one of the most important things we do."

(DV) [Richard, I think it the evidence is that cooking does increase the amount of energy that we modern humans obtain from a significant amount of our food.]

(RW) "Likewise, maybe I am wrong that humans are biologically adapted to eating food cooked. But again, that's where the evidence points."

(DV) [Richard, no it doesn't – at best the evidence does not conclusively contradict your pet theory – which is not the same thing as the evidence pointing to your pet theory being correct…]

(RW) "If you know of evidence that contradicts that position, let's hear it!"

(DV) [Richard, haven't I done that in my book? Surely you don't want me to repeat it all again here!... Death by filibuster?!?... Crucially, raw foods, in various sensible proportions by food group, with the use of primitive tools for digging, cracking, grinding and mashing, and drying for storage, and rehydration via soaking, as required, can meet the requirements! There is no need for cooking to meet energetic requirements; there is much better evidence for early tool use than there is for early fire use…]

(RW) "There will be intriguing puzzles to come, such as how humans ate with their small teeth and guts in the Lower Paleolithic if they were not using fire."

(DV) [Richard, the teeth and guts aren't very small, only a little so – rather than fire they more likely used primitive tools for digging, cracking, grinding and mashing, and drying for storage, and rehydration via soaking, as required, to eat suitable proportions of (raw) nuts, oily seeds, cereal grains, other seeds, (correction: and legumes) starchy root vegetables and fruit, along with some amount of leafy green vegetables…]

(DV) [Richard as before, any response you make to my responses here will be included verbatim and in total, no matter how negative, at the end of my book.]

Professor Wrangham's response on 18 November 2010 to the above was:

> Dear David,
> Thanks for this. It is interesting to see your perspectives. We
could
> doubtless continue the discussion for a long time.
> Can I mention one point. You seem to think that I am committed
in some
> personal way to carnivory (and to cooking!). I do not feel in the
> least committed to either. Although I am not a full vegetarian (I
eat
> fish and eggs) I do not like to eat animals that I would not be
> personally prepared to kill, and I have not eaten mammals since
1977.
> I will move on at this stage, and wish you luck in writing your
book.
> Richard

And my immediate response to that (other than additions made after the fact as "(abc … xyz)" was:

Hello Richard,

I am pleased to hear your ethics regards eating animals are better than the impression you might forgive me for forming/ receiving as a result of the frequent sensationalization and exaggeration of chimpanzee carnivory - yes it was newsworthy on discovery, and worthy of study, but you will probably understand my annoyance at every last drop of publication potential being improperly milked out of it, with some very very negative effects on animal welfare resulting.
I thank you for, and acknowledge those ethics.

If you do care to respond to my communication you should discard the previous one and use the one attached here (this is what appears above here), as I have corrected a mistake made in haste (the omission of legumes from the list of food types) and improved something else (the deletion of a mistake in reasoning of mine, and its replacement with the list of colubus characteristics that should have been addressed in the study of chimpanzee hunting which I specifically criticize here above).

Yours in science,

David Vance.

Right of reply offer to Professor Wrangham, and any reply provided by him:

The author of the book critically reviewed here, Professor Richard Wrangham, will be provided with a copy of this work in May 2011, and any response he provides at any time thereafter, including an indication of intent to respond at some stated or unstated time in the future, will be added here (this particularly refers to this document/ book part as being loaded onto and available from the Multifactor Health website), verbatim as provided, no matter how negative, immediately below within 2 weeks of my receiving it. Any response from me, here, to such response from him will likewise be communicated to him within 2 weeks of my making it. And any response from him to that response, likewise. And so on until all novel points, rebuttals, arguments or theories have been included accordingly here.

Prof Wrangham's response (awaited), :

My response to that (to follow on receipt of Prof. Wrangham's response):

(Pending) Appendix 2 Dialogue with Prof. Wrangham on this detailed review of Catching Fire:

(See the Multifactor Health website for any such)

Made in the USA
Monee, IL
04 November 2026

67383426R00167